FUNDAMENTALS OF CIVIL ENGINEERING

An Introduction to the
ASCE Body of Knowledge

D0209471

FUNDAMENTALS OF CIVIL ENGINEERING

An Introduction to the ASCE Body of Knowledge

Richard H. McCuen
Edna Z. Ezzell
Melanie K. Wong

With a Foreword by David Mongan

CRC Press is an imprint of the
Taylor & Francis Group, an **informa** business

CRC Press
Taylor & Francis Group
6000 Broken Sound Parkway NW, Suite 300
Boca Raton, FL 33487-2742

© 2011 by Taylor and Francis Group, LLC
CRC Press is an imprint of Taylor & Francis Group, an Informa business

No claim to original U.S. Government works

International Standard Book Number: 978-1-4398-5148-7 (Paperback)

Visit the Taylor & Francis Web site at
http://www.taylorandfrancis.com

and the CRC Press Web site at
http://www.crcpress.com

Contents

Foreword

Most definitions of a learned profession state that it must include at least three elements: (1) an association or organization for the profession, (2) an ethical code of conduct, and (3) a definitive statement of the knowledge required to practice the profession. The civil engineering profession has had an organization, the American Society of Civil Engineers, since 1852 and a code of ethics since 1914. It is only with the publication of the *Body of Knowledge* (BOK), first in 2005 and the second edition in 2008, that the civil engineering profession defined the knowledge, skills, and attitudes needed to practice civil engineering at the professional level. Prior to then, the defining of required knowledge was left to ABET through its accreditation process.

In 1995, the American Society of Civil Engineers (ASCE) convened a summit on civil engineering education. One of the outcomes was the agreement that the current undergraduate degree as defined by Accreditation Board for Engineering and Technology (ABET) was inadequate for the preparation of civil engineers to practice at the professional level. Subsequently in 2004, the National Academy of Engineering published *Vision for the Engineer of 2020*, which stated that "almost all discussions of educating the engineer of 2020 presumes additions to the curriculum—more on communications, more of the social sciences, more on business and economics, more cross-cultural studies …, and so forth." ASCE, in 2007, published its own *Vision for Civil Engineering in 2025* that set forth an aspirational vision of what a civil engineer in the future should be. In order to achieve this vision, the report detailed specific knowledge, skills, and attitudes needed by the civil engineer of the future.

ASCE's BOK presented the knowledge, skills, and attitudes that civil engineers need to know, while ASCE's vision for the civil engineering profession defined why they need to possess these characteristics. This primer delves into how to obtain the necessary knowledge, skills, and attitudes. While not addressing the technical outcomes identified in the BOK, it focuses on the fourteen outcomes that are currently not fully explored in today's traditional civil engineering curriculum. Its objective is to help civil engineers understand their role in society, society's impact on them, and their own impact on society. The primer provides resources, discussions, and exercises to accomplish this. It is a very useful companion to both ASCE's vision document and the BOK, and is not just intended for educators but also students and practitioners. The latter group could easily benefit by reading this book and developing the needed skills and attitudes necessary to succeed in an ever-increasingly complex world.

David G. Mongan, PE, F ASCE
ASCE President 2007–2008

Preface

The second edition of the ASCE *Body of Knowledge* (BOK) states:

> For purposes of the civil engineering BOK, outcomes are statements that describe what individuals are expected to know and be able to do by the time of entry into the practice of Civil Engineering at the professional level in the 21st century—that is, attain licensure. Outcomes define the knowledge, skills, and attitudes that individuals acquire through appropriate formal education and prelicensure experience.[1]

It is quite likely that most civil engineering programs as they are currently structured do not fully meet this goal. The technical side of the BOK is probably addressed adequately, likely even more than adequately. However, all students who receive undergraduate degrees in civil engineering probably fail to adequately develop the full range of knowledge, attitudes, and skills suggested and implied by the BOK. Undergraduate civil engineering education would be greatly enhanced if the knowledge, skills, and attitudes (KSAs) stressed in the BOK were more formally addressed in the curriculum. This objective will be more easily accomplished if resource material is available. This primer was written as a resource for addressing some of the KSAs that are not specifically introduced in many undergraduate civil engineering programs.

This primer was developed principally as a reference for an undergraduate course where topics identified in the ASCE *Body of Knowledge* are presented. The material covered in this primer is limited to the nontechnical aspects of civil engineering. The material presented in the book for each BOK outcome is intended as an introduction rather than thorough coverage, as an entire three-credit-hour course could be devoted to the individual BOK outcomes like leadership and communication. In addition to civil engineering students, the primer could serve as a resource for those in other engineering disciplines, as many of the BOK outcomes are relevant to success in those fields. While the primer was conceived as a classroom resource, it would certainly be of value to those who have completed their formal education but have an interest in adding breadth to their technical knowledge.

The goal of this primer is to introduce fundamental aspects of fourteen outcomes addressed in the second edition of the ASCE *Body of Knowledge*. Having an appreciation for these topics will lead to a broader perspective and understanding of the role that civil engineers play in society, as well as the impact society has on civil engineering and the impact of engineering on the world. The following specific objectives will help the reader meet the stated goal:

1. To encourage introspection, self-evaluation, and development of a plan for increasing one's breadth of knowledge
2. To develop attitudes that are essential to achieving both one's potential and success in his or her engineering career

3. To advance the reader's professional maturity and ability to be a leader in the civil engineering profession
4. To more fully appreciate the human values that are fundamental to professional practice and understand why these values are important in engineering practice

The ASCE *Body of Knowledge* addresses the KSAs that should be mastered by engineers. Ideally, studying the material covered in this primer should encourage readers to expand both their knowledge of and interest in these topics. Ideally, readers will want to continue their professional development through study of all of the topics addressed by the *Body of Knowledge*. Discussion questions and group activities are included at the end of each chapter to invoke further introspection and research about the individual topics. Group discussions, especially if a knowledgeable facilitator is present, will provide a broader understanding of the importance of each topic to those practicing civil engineering.

REFERENCE

1. ASCE. 2008. *Civil engineering body of knowledge for the 21st century.* Reston, VA. ASCE Press.

1 Introduction

1.1 INTRODUCTION

Many practicing engineers believe that civil engineering graduates are overqualified technically but upon graduation are underqualified professionally. The BSCE graduate adequately understands and can apply the technical concepts of civil engineering, but he or she often lacks an understanding of important elements of professionalism. Common beliefs are that recent graduates are incapable of working on a team, cannot adequately communicate with clients, and lack a fundamental knowledge of the requirements to be a leader. Additional weaknesses that are often cited include a lack of understanding of the historical aspects of engineering and how engineers influenced society, the proper way to handle ethical dilemmas, an attitude of hard work and loyalty to the employing company, and the need for lifelong learning, i.e., self-study or continuing education. A complete education should expose students in some way to all of these professional issues, not just the technical aspects of civil engineering.

From this perspective, it appears that the CE graduate could be better prepared for professional practice. Civil engineering programs devote approximately 25% of the credit hours to nontechnical courses, which are often referred to as general education. These courses include topics such as history, the arts, anthropology, sociology, business, and language. Programs often fail to provide the guidance that would show students the importance of that part of the curriculum. Instead, students select a course based on what fits with their schedule, ensures an easy A, is known to require little effort, or repeats material that they have had elsewhere. Instead of learning creativity through a fantasy literature course or improving writing skills through a journalism class, the credit hours that could improve their professional perspective are viewed as a way that the university increases tuition income. Better guidance and maybe even greater control of the alternative course options could improve the chances that the student will be exposed to ideas related to the philosophies in the bodies of knowledge.

1.2 THE PHILOSOPHY OF A BODY OF KNOWLEDGE

The deficiency of graduates to be prepared for many aspects of professional life has led numerous organizations, including the American Society of Civil Engineers (ASCE), to develop bodies of knowledge (BOKs). These BOKs include guidance on both technical knowledge and the "soft" skills where graduates are often deficient. The inclusion of professional outcomes with the technical skills shows the need for breadth without sacrificing the technical strength of today's engineering education.

A body of knowledge is an educational prescription to ensure that upcoming professionals serve the needs of the society, the profession, the clients, and the firms and organizations that are involved. As such, BOKs are developed to achieve multiple ends, including the following:

- Individuals should not sacrifice breadth for specialization. The engineer of the twenty-first century will require an extensive background of knowledge.
- Learning does not end with one's formal education, but continues throughout one's career using both organized training and self-study.
- Individuals within the profession must be actively engaged beyond technical matters, rather than accepting a passive role in local and global affairs.
- Knowledge, skills, and attitudes are not outcomes that one possesses or does not possess; instead, individuals move through stages and must seek to attain a higher level of each. This is accomplished through formal education and training, practical experience, personal growth, and self-study.
- While outcomes may be listed as separate topics, achieving the status of a professional will require an interaction of their knowledge, skills, and attitudes.
- Civil engineers must be more than technicians; they need to be leaders of society. They will achieve such status only by being as capable in the professional attitudes and skills as they need to be in technical issues.
- The civil engineering profession must take on a global perspective. We can no longer limit our service to achieving local goals.

Bodies of knowledge identify important outcomes. However, they do not tell how these outcomes should or can be achieved, or the relative importance of each. Importance will vary with the individual and his or her responsibilities, including personal and professional duties. Importance will also vary over the duration of one's career. Technical ability is often most relevant early on, while leadership ability generally takes on greater importance as one's career progresses. All in all, recognizing the importance of all outcomes in a body of knowledge will aid the future engineer no matter where his or her career path may lead.

1.3 BODIES OF KNOWLEDGE AND CAREER GROWTH

It is difficult to look into one's own future and predict even the most general career paths or the outcomes of one's decisions. Therefore, it is important to prepare for an array of futures. Preparation involves obtaining a breadth of knowledge, developing an array of skills, and possessing attitudes that will enable success regardless of the professional path that one takes. Attitudes considered necessary for success include being self-confident, creative, honest, curious, having commitment and persistence, and being optimistic. In addition, the ability to communicate in a variety of professional situations is often considered to be the most important skill. Leadership abilities are also critically important to success. The ability to embrace and employ new technologies is important to both individual and organizational advancement, as this helps achieve a competitive edge. Developing the knowledge, skills, and attitudes prior to when they are needed is the purpose of both formal education and self-study. In most civil engineering curriculums, the former focuses on the technical knowledge, so individuals must develop the skills and attitudes on their own.

2 Humanities

CHAPTER OBJECTIVES

- Demonstrate the importance of the humanities in the professional practice of engineering
- Discuss the creation and evaluation of new knowledge in the humanities
- Discuss values relevant to the humanities and their importance to civil engineering practice

2.1 INTRODUCTION

The engineer must have both technical and societal knowledge to be a socially responsible citizen and a professional. As such, it is important for the engineer to be liberally educated in the humanities and social sciences, as well as in mathematics and natural sciences. Civil engineers of the twenty-first century must be critical thinkers who can analyze problems and produce creative solutions. They must also be viewed by the public as unselfish and honest, which will maintain the positive public image that civil engineers now enjoy compared with the reputations of some other groups. Civil engineers of the twenty-first century must be good communicators in order to meet their responsibilities to clients and the public, and to efficiently interact within design teams. Knowledge of the humanities helps prepare engineers to ask the right questions, to be open-minded and creative, and to communicate well to bring about solutions to the broad-based problems that confront society and the civil engineering profession. The humanities are an essential part of undergraduate curricula, yet many students do not recognize the importance of the humanities to their future as a professional.

The humanities are branches of knowledge that address human culture and include disciplines such as history, language, philosophy, the fine arts, literature, and architecture. Working definitions of these are as follows:

- *History*: The branch of knowledge that records and analyzes past events.
- *Language*: The means of communicating thoughts, feelings, meaning, or intent, with a special emphasis on transmitting knowledge of a culture.
- *Philosophy*: A system of inquiry into the nature of beliefs and values based on logical reasoning rather than empirical investigation and evidence.
- *Fine arts*: Creative works intended to invoke contemplative delight or thought rather than for utilization.

- *Literature*: Communications, usually written and often imaginative, produced by learned scholars for transmitting ideas.
- *Architecture*: The art and science of design, usually structures, for orderly proportioning.

Rather than utility, which is a central focus of courses in engineering, these definitions emphasize the human elements of society, including creativity, feelings, ideas, emotions, and aesthetics. Imaginative expression is important to engineering design. Therefore, knowledge of the humanities with its emphasis on creative thought complements the emphasis placed on utility in the engineering and science subjects. The public wants bridges and buildings that are aesthetically pleasing, not just functional. Without some knowledge of the humanities, design engineers may place too much emphasis on utility rather than on aesthetics, thus producing a skyline that fails to be aesthetically pleasing.

Engineers of the twenty-first century must recognize that a complete design should acknowledge the aesthetic, ethical, and historical considerations that are involved in making an engineering design complete. It is inadequate to view design only as the completion of the technical computations, i.e., the computation of forces and moments, of stresses and strains. The design engineer needs to understand and appreciate the benefits of a design that reflects the culture and goals of that society. Also, the civil engineer must recognize how engineering impacts society and how society impacts engineering. Knowledge of the humanities is essential for a professional to meet his or her responsibilities to society.

The ideas transmitted in this chapter are intended to show the personal and professional breadth that can be gained through thoughtful pursuit of a strong background in the humanities, which can be gained during one's undergraduate career or through lifelong learning activities.

2.2   VALUES FROM THE HUMANITIES

Values are inherent to the humanities, yet they are rarely discussed in civil engineering curriculum. Three values that are important to the public but are often underappreciated by engineers are

- *Aesthetics*: Perception of excellence in craftsmanship; beauty.
- *Variety*: Having a broader perspective; acknowledging diversity of perceptions of excellence in appearance or form.
- *Enjoyment*: The sensation of experiencing pleasure or being gratified for something done well.

Communities often take pride when local infrastructure is recognized as being aesthetically pleasing and can bring notoriety to its surroundings. People also enjoy variety and history, so they take pride in bridges of the past even though these classic stone-and-steel structures lack an artsy, modern look. Communities enjoy bridges and buildings that fit in with the surroundings or complement historical structures or artifacts for which the locality is known. In addition, communities are often willing to provide additional funds for engineered facilities that add beauty to the surroundings.

2.3 PHILOSOPHY AND DECISION MAKING

The study of the humanities develops critical thinking techniques that allow engineers to interpret information; raise the right questions; and examine the assumptions, implications, and consequences of engineering decision alternatives. *Philosophy* is the use of reasoned argument techniques to examine the nature, scope, and limits of existence, knowledge, and morals. For example, the study of Aristotle and Plato's philosophies focuses on the importance of logical reasoning in decision making. The Socratic method is a teaching technique in which philosophical inquiry is used to examine the implications of an idea and to bring about a solution. It centers about questioning of the basics of the problem. The use of the Socratic method forces students to examine every implication of a statement made and to think critically when making an argument. These examination and critical thinking skills are important for design engineers, as engineers should ask questions of their designs. Inquiring about risks and uncertainties, safety issues, and the sustainability of the project can lead to better decisions.

Engineers are confronted by the ethical dilemmas that require well-reasoned decisions. Ethical decision making requires defining the moral dilemma, developing alternative solutions, obtaining relevant information about each alternative, evaluating the alternatives, and implementing the selected alternative. While a philosophy course may use this decision process for an issue such as abortion or gun control, the decision process itself is an important educational objective in civil engineering. Seeing the generality of the process through discussion within a specific philosophical context is a better learning mechanism than seeing the process applied solely to a decision about managing a construction project. Understanding the general process will enable a person to apply it to a broad array of problems. Philosophy also encourages examination of personal values and morals, which ensures that an engineer will make decisions for the common good, especially when he or she is challenged by competing influences or objectives.

A primary responsibility of leaders is to make decisions. Leaders in civil engineering design firms make decisions on a regular basis, including personnel selection, whether or not to bid on a proposed project, which piece of software is most appropriate for a particular design, and establishing organizational goals for the future. While we often associate decision making with business management, it also falls in the realm of philosophy, which is a primary part of humanities. The process used by philosophers to make decisions about moral issues is quite similar to the decision process used by both business managers and leaders of engineering firms. Thus, a philosophy course in moral decision making is relevant to the engineering student from a professional standpoint and even more from a personal perspective.

2.4 ART AND CIVIL ENGINEERING DESIGN

While an understanding of the fine arts can be personally rewarding, it can also improve an engineer's ability to design effectively. The fine arts include sculpture, painting, drawing, architecture, literature, drama, music, and dance. Understanding the strategies applied to creating art can be applied to the design of both structures and engineering products that are aesthetically pleasing, yet functional. Knowledge of the arts can develop a person's creative ability, which is necessary to produce innovative solutions to societal problems that occur in a rapidly changing technological society. An appreciation of the arts also enhances communication skills and enables an engineer to provide the public with infrastructure that goes beyond functionality.

An engineer may view art solely as an attempt to capture reality (e.g., a bowl of fruit or a mountain landscape) into a simplified two-dimensional representation on canvas. Conversely, the artist may consider art as an attempt to communicate certain emotions or feelings with the viewer. To the artist, the person who appears in a painting standing at the base of a mountain may reflect the struggles that a person faces in life. An appreciation of art can expand one's thinking beyond a narrow utilitarian viewpoint and encourage adopting broader thinking that values emotions and feelings.

A course that discusses art appreciation would teach the engineering student to recognize the importance of balance, proportion, variety, and unity, which are characteristics that the artist uses in his or her work. These characteristics of a painting are similar to those that design engineers use. A design engineer's building is more than just lines, shapes, materials, and patterns. Likewise, a portrait or mural is more than just these characteristics.

A course in art appreciation emphasizes the following characteristics of a piece of art:

- *Balance*: A sense of visual symmetry for the sectors of the art piece.
- *Proportion*: An artist uses proportions of objects in a picture to control the feelings of the viewer by placing emphasis on certain objects within the art piece. An unrealistic proportion may be used for emphasis and seem to be injudicious by the rational viewer, but it may be the artist's way of invoking the viewer's thoughts and feelings.
- *Variety*: Variations in color, shading, and object shapes can add variety to art, which can invoke feelings in the viewer.
- *Unity*: While variety is important to keep the artwork from being mundane, unity is necessary to ensure that the artwork as a whole is seen as a single idea or concept.

In the design of a building, the civil engineer should ensure that each of these characteristics is considered to create a design that is aesthetically pleasing yet functional. The building should provide balance with respect to other buildings in the communities, while still remaining sufficiently different to provide variety. The downtown area of a city populated by a series of rectangular buildings lacks variety, even though the buildings may be very functional. This is a very bland environment. If some buildings were designed to have shapes or proportions different from the other buildings, then the variety would likely add aesthetic pleasure to those who are walking through the neighborhood. Such places become known by stark physical differences, i.e., the circular building or the pyramid. The building with a novel shape can provide an economic boom for a neighborhood as companies want to be associated with the distinctive design. Environmental beauty can be preserved at a building site, such as the case in Frank Lloyd Wright's Fallingwater, which incorporates a stream into the design of a home. Artistic understanding can encourage the inclusion of socially desirable characteristics in engineering designs.

Taking a humanities course in art has both personal and professional benefits.

An art course will be a better means of learning these characteristics than a few side remarks made in a structural engineering course. When taking an art appreciation class, the engineering student may simultaneously learn to enjoy a trip to a local art museum and to consider characteristics like variety and proportion in the design of buildings.

2.5 THE IMPORTANCE OF LANGUAGE TO A PROFESSIONAL

The first impression that someone might have when the issue of language is raised is the importance of knowing a foreign language. Obviously, an engineer who knows a foreign language may have more opportunities for assignments in foreign countries. However, the importance of language goes well beyond the study of foreign languages. The definition given at the beginning of this chapter viewed language in a much broader context. Specifically, language is the means of communicating thoughts and feelings. Language helps to persuade a client that your firm is the best one to complete the job, express reasons why your design solution is best, refute irrational reasons of competitors, motivate subordinates to work toward organization goals, and most obvious, make oral and written communications more effective. To a professional and especially to a leader, persuasion, motivation, and transmitting knowledge are extremely important elements of language.

The study of language involves both grammar and vocabulary, both of which can be applied to learning a foreign language. Understanding a foreign language, like the Tzeltal language of the Maya in Mexico, which has twenty-five words for the idea "carry," provides perspective to how important carrying was to the Mayan society. In the same way, foreign language studies can provide understanding of the differences in the way people categorize their experiences. Facility in a foreign language can also allow work in a globalized environment and the possibility for a design career in a foreign country. Foreign language studies can also enhance understanding of other cultures and business relations with foreign businesses.

Does someone with exceptional persuasive skills have an advantage over someone from another firm who is not persuasive in marketing his or her company? *Persuasion* is the process of changing the attitudes, behavior, or beliefs of another person through the use of language. The benefits of being skilled at persuasion are many, both in one's personal and professional life. Persuading a procrastinator to complete his or her part on a team project will enhance a person's reputation as a leader. Persuading a potential employer that you are the right one for the job will help you get the job that you want. Language skills are central to persuasion, and someone who lacks the ability to persuade is unlikely to rise to a leadership position.

CASE STUDY

Employee efficiency, which is important to the success of a company, can depend on the language skills of leaders. One role of a leader in a company, from the CEO to a project manager, is to motivate employees to complete their work with excellence and in a timely manner. Subordinate motivation depends to a large extent on the ability of the manager to orally persuade subordinates that task completion is important. But what if the manager uses language that causes the subordinate to feel overly stressed? Then does persuasion and motivation have efficiency implications? Good language skills will likely lead to timely, quality work without extraneous amounts of stress. Thus, a lack of good language skills can be responsible for somewhat inefficient behavior.

2.6   HUMANITIES AND A CULTURAL PERSPECTIVE

The word *culture* is often associated with the study of a historical group, such as the Mayans, a primitive group on some South Pacific island investigated by Margaret Mead, or upper-class people who visit art museums and attend Mozart and Beethoven performances. In reality, culture is the socially transmitted behavior patterns, beliefs, and institutions of a community. While this does apply to the Mayans, the word *culture* also applies to the engineering profession, which involves both beliefs (e.g., public safety is important) and institutions (e.g., licensure). The study of other cultures can provide a perspective on the engineering profession and its role in the community.

The study of humanities allows the civil engineer to understand his or her own professional culture. An understanding of the culture in which an engineer designs is essential to how the design meets the needs of society. Courses in the humanities provide perspectives on how a design will be used by the client or community and the impacts that a design can have upon society. The study of literature involves the analysis of the thoughts embedded in the literary works and their societal implications. Religious studies courses can provide perspective on the values of others through the study of the history, moral principles, and interconnectedness of different religions. A general course on world religions can provide knowledge of the morals and values of people in other countries that may impact their decisions in engineering design and their attitudes in the workplace. This is especially important as civil engineering integrates itself globally. Understanding culture through philosophy, literature, and religious studies allows engineers to understand how they should approach design and the scope of the design's impact on society.

In the last century, new technologies have impacted society in ways that earlier engineers could not have foreseen. Quality of life is constantly improving as greater efficiency in manufacturing and agricultural practices allows for cheaper products, a more varied diet, and more leisure time. Safety improvements in structural design have allowed for greater peace of mind due to lower risks of infrastructure failure, while better heating, cooling, water, and sewage systems have improved the standard of living. Yet, all of the global society has not reaped the benefits of this technological progress.

Reading novels can provide insight into the ways that machines change society and the value conflicts that may confront engineering management in the future. The advent of mechanization during the industrial revolution negatively impacted some aspects of society. During the industrial revolution, many skilled workers lost their jobs, while job satisfaction declined as most of the jobs required completing monotonous tasks. Kurt Vonnegut's novel, *Player Piano*, takes place in a mechanized society and details the dilemma of an engineer who must fire laborers as machines replace human skills. When reading fiction, it is important to view the events and characters in a broader context, especially considering the potential application to the engineering culture.

2.7 HUMANITIES RELEVANT TO PERSONAL AND PROFESSIONAL DEVELOPMENT

The word *development* is generally applied by psychologists to the maturation of young children. Herein, the word *development* will be applied in both personal and professional contexts. The term *professional development* refers to maturational changes in behavior that occur in a professional setting. The term *personal development* applies to maturational changes relevant to personal behavior. The two are not independent, as maturation in one is often associated with maturation in the other setting. For example, a person who finally recognizes that a particular action is selfish and decides to change may revise his or her selfish behavior when acting in either personal or professional settings.

It is essential to be liberally educated to fulfill the requirements of the professional civil engineering discipline. A liberal education allows the engineer to understand the relationship between society and professional engineering, and furthers personal development. A study of the humanities improves self-awareness, self-confidence, and communication skills. A liberally educated civil engineer will be more socially sensitive to future developments and contribute a well-rounded perspective to the engineering profession.

The humanities offer opportunities for personal development. A person who lacks leadership experience and the confidence to pursue a position of leadership would benefit from a history course that discusses why leaders were successful. Even after being burned at the stake because of the accusation that she was a witch, memory of the woman still motivated her followers. History provides many examples of leaders such as George Washington, Winston Churchill, and Henry Ford. Understanding the qualities that enabled them to be successful can provide guidelines for a young engineer to overcome a lack of confidence and become a leader.

For a person who lacks a broad sense of self-confidence, courses in the fine arts and literature can be of value. For example, a course that would have the person write a piece of fiction can improve his or her ability to express ideas and think without constraint or fear of communication. Similarly, a course where a physical experience of art, i.e., sculpting or painting, would allow the person to act freely, less constrained, can help him or her overcome a lack of confidence and be less hesitant to communicate orally.

Selfishness runs counter to developing a solution to many problems, in both personal and professional life. Activities such as Engineers Without Borders provide the opportunity to experience the importance of selflessness and an altruistic philosophy. A course in philosophy that deals with value can encourage reflection on personal development away from selfishness. Other courses in philosophy deal with materialism and its implications, and the broader conception of nature. Such a course might develop an awareness of aesthetics and feeling for sustaining biota. A study of Aristotle or Spinoza will help a person view man as part of nature, not those who are free to dominate nature. Recognizing the importance of selflessness can lead to personal and professional growth away from selfishness and to decisions based on a more robust view.

2.8   ROLE OF CURIOSITY IN ADVANCEMENT

Curiosity, which is an important attitude for civil engineers (see Chapter 13), has historically played a significant role in the advancement of knowledge, and future advancements in civil engineering design methods will depend, in part, on our curiosity toward problems that arise. The study of the humanities can illustrate the professional value of curiosity, as curiosity is an important attitude in the advancement of new knowledge.

Curiosity has played a significant role in the evolution of dance. Dance is a communication of emotions and a creation of visual designs that involve movements. It is often choreographed in a way that emphasizes new, unique variations. It was the curiosity of choreographers such as Isadore Duncan and Martha Graham in the early twentieth century that enabled modern dance to become popular.

The benefits of curiosity are also evident in the history of science and engineering. For example, Sir Isaac Newton held a position equivalent to a modern-day college professor when he completed many of his advancements in mathematics. However, he was not encouraged to advance knowledge as part of his responsibilities as a professor. His personal curiosity was his motivational force, and we are the beneficiaries of his curiosity.

William Froude (1810–1879) was curious about the resistance of ship hulls. He conducted experiments to understand the role of friction. This led to the Froude number, which is a basic concept used in fluid mechanics and hydraulics courses.

Sadi Carnot (1796–1832) can be considered a founding father of thermodynamics because of his interest in the steam engine. He was concerned with practical aspects, such as its efficiency and differences between the workings of the ideal and actual engine.

The curiosity of individuals has been a driving force throughout history. However, during the last century, group curiosity became necessary to solve many problems. The Los Alamos group developed the A-bomb. The Wright brothers acted as a team in the development of the airplane. The Lorenz Company developed radar. In these cases, curiosity and competition combined to advance knowledge.

Human curiosity has always been a dominant force in the advancement of science and engineering, and we can expect this to be true in centuries to come. For the profession to solve the problems of the future, civil engineers will need to develop new knowledge. This requires curiosity, the attitude of wanting to know and learn about problems and their solutions. An appreciation of the humanities can enhance this attitude.

2.9   THE CREATION AND EVALUATION OF NEW KNOWLEDGE IN THE HUMANITIES

Knowledge in the humanities develops in much the same way that knowledge in civil engineering develops. A human need is recognized, a goal or hypothesis is formulated to serve as a direction for investigation, an experiment is designed and conducted to study the problem, and the analyses lead to the acceptance or the rejection of the hypothesis. The experimental process of Section 4.5 can be applied to solving problems and creating new knowledge in the humanities. Identifying the problem is generally the most difficult part of the process. Very often, the problems are identified by society as needs develop; however, individuals can identify problems that require study through rational thought. A few questions related to new knowledge in the humanities are as follows:

- *History*: What was the public's perception of risk about traveling on public transportation during the era of steam boiler explosions (early 1800s)? How does it differ from the perception of risk now?
- *Language*: Does text messaging have an effect on language development?
- *Philosophy*: Has increased environmental concern changed the concept of nature?
- *Fine arts*: Is obesity in the United States changing the style of dance?
- *Literature*: Has the issue of global climate change influenced the direction of literature in the early twenty-first century?
- *Architecture*: Has the threat of terrorism influenced architectural design?

When civil engineering students enroll in humanities courses, they should practice using their creative powers to identify new knowledge. Practicing this skill will also be of value in advancing knowledge of civil engineering issues.

New knowledge, whether it is related to the humanities or engineering, must be evaluated to ascertain the accuracy, importance, merit, or benefit of the new knowledge. The new knowledge must be logical and reasonable on the basis of observation and thought, in order to be valid. Valid statements of new knowledge will resist challenge. The concluding statements that reflect new knowledge should show consistency of reasoning. Judgments should be unbiased.

2.10   OBSERVATION

The sections of this chapter have shown connections between individual branches of humanities and ideas relevant to the practicing civil engineer, e.g., philosophy and decision making. The engineer must recognize that the value of the humanities goes well beyond these illustrations. For example, a knowledge of philosophy would enable a civil engineer to appreciate the aesthetic value of a wetland; facilitate research about new materials in civil engineering design; and use rationalism and empiricism in interpreting observations and the results of experimental studies. Of course, knowledge of the humanities can also lead to personal enjoyment.

2.11   DISCUSSION QUESTIONS

1. Identify subdisciplines of history. For each one, provide a one-sentence description.
2. Identify subdisciplines of language. For each one, provide a one-sentence description.
3. Identify subsdisciplines of philosophy. For each one, provide a one-sentence description.
4. Identify subdivisions of the fine arts. For each one, provide a one-sentence description.
5. Identify subdivisions of literature. For each one, provide a one-sentence description.
6. Discuss how knowledge of the industrial revolution is relevant to the civil engineer of the twenty-first century.
7. Discuss how knowledge of the history of technology relates to problems that the civil engineer of the twenty-first century faces.
8. Discuss how knowledge of language arts can be useful to an engineering leader.
9. Discuss the benefits, personal and professional, of knowing a foreign language.
10. Discuss the importance of persuasion in teamwork.
11. Discuss the role of language in motivating subordinates.
12. Discuss the role of language in persuading a client to award the job to your employer.
13. Discuss the relevance of knowledge of axiology to a design engineer.
14. Explain the relationship of epistemology and lifelong learning to a civil engineer.
15. Discuss how a course in sculpture could benefit an engineer.
16. Discuss the importance of aesthetics to engineers and how an engineer can learn about aesthetics from an art appreciation course.
17. Explain how a course in fantasy literature could benefit an engineer.
18. Examine the song "Little Boxes" by Malvina Reynolds, and how its perception of suburbia can be applied to creating housing developments that are socially desirable.
19. Examine E. M. Forster's *A Passage to India* and how his insight to an Eastern perspective can be important when practicing professional engineering in India.
20. Discuss how the design of the National Museum of the American Indian in Washington, D.C., employs the principles of artistic design and how it considers American Indian culture.

21. Four artists may paint the same mountain, yet the pictures might be quite different. The style of art can be identified by four key concepts related to style: emotion, fantasy, order, and accuracy. Discuss how the mountain might appear in a picture when each of the four styles is dominant.

2.12 GROUP ACTIVITIES

1. Demonstrate the importance of the humanities to engineers who are in leadership positions.
2. An employee of an engineering company is threatening to blow the whistle on the activities of another employer. Identify ways that knowledge of the humanities can help the manager, a civil engineer, resolve the issue internally.

3 Social Sciences

CHAPTER OBJECTIVES

- Understand the importance of knowledge of social science to the civil engineer
- Analyze civil engineering problems where social science issues are relevant
- Create and evaluate new knowledge in the social sciences

3.1 INTRODUCTION

- How do people collectively react during a natural disaster like a hurricane? In what type of course would a civil engineering student learn about the behavior of masses?
- Why are some individuals confident while others are fearful of interacting with others? In what type of course could you learn about attitude development or change?
- Is economic growth and environmental preservation a zero-sum game? Or can economics provide insight into ways of curbing environmental destruction? What type of course would deal with this apparent economic versus environmental conflict?
- Does civil engineering benefit from social research or even research in general? Does the federal government have a responsibility to support research from which private companies will benefit? What type of course would discuss the federal role in research?

Should civil engineers be concerned about human behavior? Attitude development? The environment versus economics quandary? Research? If yes, then social science courses in sociology, psychology, economics, and political science are of professional and personal benefit.

The civil engineer of today should seek knowledge from the social sciences, which will benefit a civil engineer in both his or her interpersonal relationships and professional career. Disciplines in the social sciences include sociology, psychology, economics, political science, anthropology, history, geography, and economics. Knowledge from the social sciences can contribute to team management, teaching and learning, preparing and dealing with natural disasters, land use planning, mass transportation design, consumer markets, risk analysis, and environmental solutions.

This chapter presents concepts from several social science disciplines and their relationship to problem solving in civil engineering.

3.2 DEFINITION: SOCIAL SCIENCES

Generally, *social science* refers to the study of society, including individual or group relationships. *Sociology* is the systematic study of personal and global social relationships and cultures. *Anthropology* is similar to sociology, but is associated with the study of primitive cultures and with a broader historical and geographic space than sociology. *Psychology* is the study of human mental processes and behavior. The study of these disciplines can develop an understanding of human behavior and interactions. Women's studies courses can also provide perspective on the roles of women in society. In summary, the following disciplines are generally considered social sciences:

- *Sociology*: The study of human social behavior.
- *Psychology*: The science of behavior, including the emotional and behavioral characteristics of individuals and groups.
- *Economics*: The management of materials, personnel, or business activities.
- *Political science*: The study of government processes, principles, and structures of political institutions.

Anthropology and some aspects of history are also considered to be social sciences.

The importance of the social sciences to civil engineers should be evident. A few specific relationships include, among others:

- Group dynamics of teams within a civil engineering company
- Human behavior in traffic accidents
- Leaders helping subordinates overcome a lack of confidence
- Planning for evacuation of burning buildings
- Helping a politician develop a public policy on environmental sustainability
- Human reactions during natural disasters
- The role of government in cleanup after a disaster
- The reactions of people during disruption of lifelines
- The movement of people in land use planning
- Consideration of risk factors in project economics

Civil engineering services are delivered to people through social mechanisms; as such, it is important to understand that the social sciences are foundational to effective service by those in the civil engineering profession. Much like engineering, the social sciences are data driven, quantitative, and analytical. However, civil engineering applies the scientific methods of the social sciences to real problems. The study of social sciences such as economics, political science, sociology, and psychology also allow civil engineers to understand how to work within a social framework and consider the nontechnical ramifications of their actions and decisions. The process of development, delivery, and evaluation of solutions that improve society are also

---

CASE STUDY

A study of the sociology of sports is relevant to engineering. Many character traits that influence success in sports are the same attitudes relevant to success in civil engineering. For example, John Wooden, the former UCLA basketball coach, whom many consider to be the greatest coach of all time, developed a pyramid of success (Wooden and Sharman, 1975). Some of the attitudes that formed the foundation of the pyramid included honesty, confidence, reliability, industriousness, team spirit, and loyalty. These are attitudes inherent in the ASCE *Body of Knowledge*. Viewing their relevance to success through the sociology of sports can provide a different but important perspective on success in civil engineering.

---

enhanced through the study of social sciences. The relevance of history to the engineer's role in the community is discussed in Chapter 2. Public policy, which can reduce risks associated with design, is discussed in Chapter 9.

3.3   INTERPERSONAL SKILLS AND THE SOCIAL SCIENCES

Knowledge of sociology, anthropology, and psychology can enhance interpersonal skills in the workplace. For example, the efficiency of a team can be improved. Civil engineers frequently work in teams on design projects, so it is essential to be able to work well with others in a team setting. Specific strategies grounded in sociology and psychology allow a team member to achieve these goals through an understanding of social relationships and social groups. For example, principles of team building, intervention into groups with personnel problems, and problems with motivation are issues addressed in courses on sociology and psychology. A perspective on others' cultures and mindsets when interacting with a diverse team is essential. As globalization becomes more embedded within civil engineering practice, more civil engineers will need to have an appreciation of worldwide cultures. As another benefit of study of the social sciences, sociology and psychology can provide insight into how to critique others effectively and understand the personal motivations of others within the spheres of peer-to-peer and subordinate-to-supervisor relationships. These disciplines can also provide the civil engineer with strategies to work effectively within alternative organizational structures, whether a corporation or small company. An understanding of varied backgrounds and attitudes allows a manager to be an effective communicator within a team or organization. A manager who is sensitive to other cultures and points of view will be better equipped to successfully handle problems and is more likely to be respected by his or her team members.

It is inevitable that at some point in a professional's career he or she will teach or mentor, whether in academia or not. As a team member or leader, the civil engineer will need to impart knowledge or expertise in a subject about which others may not have a solid background. Psychology provides an understanding of learning strategies that can improve the ability to teach others. Some may learn better through

---

CASE STUDY

Mentoring should be more than discussions of office politics and professional development activities. The organizational culture should be discussed, which includes the values and norms of the organization. A psychology course that discusses emotional intelligence would be relevant to mentoring in engineering. Ashforth and Saks (2002) identify three elements of emotional competency:

- Understand the emotional culture of the organization.
- Reflect on one's own emotional abilities and deficiencies.
- Accurately appraise the emotional intelligence of subordinates.

These are relevant to the mentee's success within the organization and his or her rise to a position of leadership.

---

reading, while others may learn through action. It is important to understand that everyone may not learn in the same way, so a mentor or teacher needs to adjust his or her strategy accordingly.

3.4 PHYSICAL GEOGRAPHY AND DESIGN FOR NATURAL DISASTERS

Geography is often naively viewed as learning the names of the state capitals and understanding the processes related to the formation of mountains. Geography involves much more than these narrow views, and the study of geography has important personal and professional benefits. Civil engineers regularly deal with the effects of natural disasters, whether it is the cleanup following an event or the design of protective measures that will minimize damage when an event occurs. Levees to control flooding, debris dams to store mudslide material, and in-stream structures to prevent ice jams in bridge openings are a few of the preventative measures that civil engineers design. Each of these designs will be better understood if the design engineer understands the relevant principles of physical geography. For example, understanding landforms and how they are shaped by flowing water is a topic covered in physical geography. This understanding is central to appreciating how mudslides occur and how the debris material is transported to low-sloped areas where it is deposited. This understanding is important to the civil engineer responsible for designing a debris dam. These aspects of landforms would not likely be covered in a typical civil engineering course.

A physical geography course also addresses climate classification and characteristics. It is generally accepted that global climate change will impact civil engineering design. Therefore, regardless of their specialization within civil engineering, knowledge of climate can be of value to civil engineers. Some studies suggest that global climate change will increase extreme climate conditions, which could increase the occurrence of natural disasters. Physical geography could also be of value to someone interested in sustainability. This course could discuss natural resources; hydro-electric, wind, and geothermal power; and soil erosion and conservation.

3.5   ENGINEERING AND ECONOMIC IMPACT

Civil engineering is a people-serving profession that also involves business transactions. Therefore, the practice of civil engineering requires knowledge of economics. As an engineer, the evaluation of the economic viability of a project or design is important. A new design of a product can create an entirely new market, and thus impact the economy. The automobile created markets for transportation systems, auto repair services, and insurance services, and spurred the development of the petroleum, iron, steel, and rubber industries. As another example, civil engineers recognized the potential value of software to the design engineer, and some created businesses to serve this need. An early exposure to the business side of the engineering practice will encourage practicing engineers to take advantage of business opportunities and provide engineering services that the design engineer requires.

Economics is often an important decision factor in engineering projects. However, other quantitative and qualitative criteria enter into decisions. Of course, all of the decision criteria are uncertain, and the degree of uncertainty adds to the complexity of decision making. Costs associated with labor and materials may be a big factor in the decision-making process, but the uncertainty of other factors, such as environmental effects and the need to sustain traffic flow during construction, may weigh more heavily in selecting among alternative proposals. Changes in the cost of energy can also influence the decision process. Learning about first costs, fixed costs, variable costs, marginal costs, and sunk costs is important to project evaluation. General economic principles studied in economics and business courses are of primary importance to practicing civil engineers.

Risk and uncertainty (see Chapter 7) are central to economic decision making. Weather conditions, labor problems, machine failures, theft, and collapses are just a few risk-related factors that must be considered in economic analyses of projects. Risk assessments must address factors such as

- What potential hazard factors are possible?
- What losses will accrue if any of the hazard factors occur?
- What is the probability that the hazard factor will occur?
- How can the risks be reduced?

While some of these hazard factors can be analyzed with some degree of certainty (e.g., flood damage), others are more difficult to address in economic decision making. Knowledge of economics should enable the practicing civil engineer to better assess the economic effects of risks and uncertainty.

3.6   REGIONAL ECONOMICS, LAND USE, AND TRANSPORTATION PLANNING

Resource economics, which is a specialty within the broad field of economics, deals with the location patterns of people and industries and how patterns of development influence profits. Location is relevant to profit. Just imagine if three McDonald's franchises were located on the same block in a suburb. It is doubtful that they could each generate a profit because their locations were not judiciously selected based on fundamental principles of regional economics.

Cities grew exponentially during the industrial revolution. Even today in developing countries, cities are experiencing exponential growth. However, cities are not located randomly. Instead, as principles of regional economics would dictate, cities grew where the necessary natural resources and amenities needed to support people and businesses were available. In the mid-twentieth century, Pittsburgh was known as Steel City because steel production was its main industry. Pittsburgh is located on the Ohio River, which provided transportation to bring the raw materials into the steel mills and then transport the finished products to the markets in the Midwest. Pennsylvania also had considerable coal deposits that encouraged economic growth. Location theory states that the location of natural resources and markets are two important factors for long-term economic growth. Similarly, the city of Baltimore grew partly because its port facilities made it a regional hub for imports and exports. Atlanta is an air traffic center because of its central location to the entire southeastern United States. The study of one aspect of economics, namely, regional economics, has important implications for the civil engineer.

The civil engineering student who takes a course in regional economics will better understand the forces related to land use and transportation planning. At the same time, the student will learn how to apply economic concepts during projects related to location, labor and capital migration, and public policies related to urban growth. Many of the same principles apply to the distribution and use of natural resources, which relate to sustainability and environmental management.

CASE STUDY

Both civil engineers and those in the social sciences have an interest in urban growth. In 1939, a sociologist, Homer Hoyt, recognized that the new modes of transportation were changing the nature of city expansion. Prior to that, the patterns of expansion were primarily concentric circles. Hoyt recognized that expansion was then beginning to occur along a web-like pattern that followed new highways and mass transit routes. Author and sociologist, Jane Jacobs, recognized that land uses were a factor in the decay of cities. She cited the expansion of parking areas as an instigator of decay and cause of flight from urban areas. The sociological effects of the works of engineers on society can often be recognized by sociologists as factors that determine patterns of urban growth and decay.

3.7   ANALYSIS OF ENGINEERING PROBLEMS WITH SOCIAL SCIENCE IMPLICATIONS

3.7.1   NATURAL DISASTERS

A study of the social sciences can provide a perspective on the preparation for and cleanup following a natural disaster. Knowledge of psychology can provide an understanding of people's psychological reactions to natural disasters and how to prepare people effectively for an event without inciting hysteria or panic. An understanding of how communities react to a natural disaster can be used to mobilize relief efforts. Civil engineers are heavily involved in flood control and often have responsibilities related to risk communication; specifically, they inform the public about changes in the array of possible flood levels, the likelihood of levee failures, and the best evacuation routes. Knowing something about the psychological nature of people can help engineers better communicate about flood risk. An engineer's knowledge of economics can help prepare a community to deal with cleanup preparation. Economic analyses are necessary in setting the heights of flood control levees. An engineer who understands the political processes of a community is better able to assist in public policy formulation. An engineering student can learn about public policy and the political process by electing to take courses in political science. Thus, the social sciences are important to engineers who deal with many aspects of natural disasters.

3.7.2   ENVIRONMENTAL ISSUES

Environmental pollution is a problem that requires a multidisciplinary approach to achieve a realistic solution. The engineer must work closely with others from an array of technical disciplines, including geology, soil science, and natural resources, as well as social science. Knowledge of economics can provide a perspective as to why communities do not embrace recycling, possibly because of debilitating initial costs or because it is not profitable enough in the short run. Regional economics can help communities determine the best location for a recycling center. An understanding of political science may reveal that the current policies or institutions are helping or hindering the reduction of environmental pollution. An understanding of the public policy process can help an engineer determine the best strategy that he or she can employ to make a positive impact. Psychology and sociology provide insight as to how people perceive environmental problems such as global warming, tainted water supplies, and ecological habitat destruction. These disciplines can be engaged to create a public awareness campaign that would effectively create interest and concern in these issues.

3.7.3   REMOVAL OF DAMS

After the Great Depression, the Tennessee Valley Authority was created to provide jobs and hydroelectricity to the Tennessee Valley through the building of dams. Now, we realize that dams can be harmful to the environment. Problems such as fish spawning and sediment transport have led to the removal of some dams. However, it is important to consider the social impacts of such actions. The Tennessee Valley

Authority was created to rapidly develop the area through the creation of new jobs and to provide inexpensive energy to the area. However, the removal of the dams now may damage the area economically by eliminating jobs and requiring residents to rely on more expensive energy sources. Hydropower is considered a renewable resource. Special interest groups that may have the interests of the environment or the businesses affected by the removal of the dams may get involved in the political process and ultimately hinder action that would improve the environment. Thus, dam removal has environmental impacts, economic consequences, and social implications. These issues can be addressed in social science courses.

3.8   THE CREATION OF NEW SOCIAL SCIENCE KNOWLEDGE

The question "What is knowledge?" is best left to the philosophers. A dictionary definition would use terms like *specific information* or *verified understanding*. Knowledge may be gained through rational thinking. Believing that something is true is not adequate justification for specific information to be true, so knowledge is not justified solely by believing. Knowledge is justified as true through experimentation and good scientific reasoning. Only knowledge that can be used is knowledge that is socially powerful.

The advancement of knowledge involves recognizing a problem, proposing a hypothesis, studying the variables involved, and then developing the experiments and doing the rational thinking needed as justification. As civil engineering is a people-oriented profession and sociology involves human social behavior, developing new knowledge about the social sciences is relevant to the practice of civil engineering. Consider the following ideas for social science research:

- *Sociology*: Do constructed terrorist barriers make people feel safe?
- *Psychology*: What is the effect on feelings of job satisfaction when a manager of engineering projects who uses a democratic style of leadership is replaced with a manager who uses an autocratic style?
- *Economics*: What is the economic cost of failed risk communication messages?
- *Political science*: What factors influence the effectiveness of public policies written on the application of best management practices for controlling urban/suburban flooding?

Note that each of these questions has a basis in a particular social science, yet they are all relevant to civil engineering practice. Knowledge gained by experimentation and rational thinking would make the design engineer more effective in his or her practice.

Ideas for new knowledge often come from experience. The individual is generally curious and likes to understand observations. The question about terrorist barriers may arise upon seeing barriers around a building and wondering how effective they would be. The question about job satisfaction may arise in the mind of someone who has experienced a change in leadership with a drop in personal job satisfaction following the change. The question about a failed risk communication may result from seeing a TV news clip that discussed the potential benefits that would have accrued if the communication had been more specific or timely. The public policy question may arise in the mind of someone who is reviewing design standards that are significantly different from those of neighboring counties or states. Answering each of these questions would be socially beneficial and of value to practicing civil engineers.

3.9 VALIDATING NEW SOCIAL SCIENCE KNOWLEDGE

Synonyms for *validity* include *sound, supportable*, and *incontestable*. A conclusion is thought to be valid if the new knowledge is based on sound premises and the conclusion can be logically derived from these premises. While statistical validity can be applied to some quantitative results, rational validity is a more important test. To ensure that the new knowledge is valid, it needs to be based on evidence, and the characteristics and the source of the evidence are considered while judging the validity.

Four characteristics of the evidence will be considered:

- *Clarity*: The evidence must be free from doubt, as it is stated.
- *Consistency*: The evidence should agree with past findings.
- *Verifiable*: The evidence would lead to the same knowledge if the study was repeated independently.
- *Sufficiency*: The quantity of the evidence is enough to ensure that the knowledge is truthful.

The quality and quantity of the evidence are important for deciding whether or not the knowledge can be considered truth.

The validity of knowledge also depends on the source of the knowledge. This includes both those who generate the knowledge and the methods used in developing the knowledge. Knowledge that resulted from poor experimental designs are inadequate and unreliable. With respect to the individual(s) who generated the knowledge, important characteristics include

- *Competency*: Having the knowledge and experience to generate the new knowledge.
- *Unbiasedness*: Not having any predisposed or prejudicial attitude to bias the findings.
- *Reliableness*: Having a reputation for honesty and dependability in past efforts of creating new knowledge.

In most cases, the person may not personally know the source of new knowledge, so the affiliation of the source of knowledge is used as a surrogate indicator.

With respect to the methods used in developing the new knowledge, evaluation criteria include

- *Approach*: The methodology used, including its complexity and whether or not it is the accepted state of the art.
- *Breadth*: The scope of the analyses, including the inclusion of both extreme conditions and minor factors.
- *Care*: The thoroughness, including the accuracy of any measurements, taken throughout the effort in developing the new knowledge.

Before the new knowledge is accepted, all of these criteria must be assessed. If the civil engineer is developing the new social science research, then he or she needs to consider each of these criteria to ensure confidence in its validity.

3.10 DISCUSSION QUESTIONS

1. Identify subdisciplines of psychology. For each one, provide a one-sentence description. Discuss its potential relevance to civil engineering.
2. Identify subdisciplines of sociology. For each one, provide a one-sentence description. Discuss its potential relevance to civil engineering.
3. Identify subdisciplines of economics. For each one, provide a one-sentence description. Discuss its potential relevance to civil engineering.
4. Identify subdisciplines of political science. For each one, provide a one-sentence description. Discuss its potential relevance to civil engineering.
5. Identify subdisciplines of anthropology. For each one, provide a one-sentence description. Discuss its potential relevance to civil engineering.
6. Discuss the relevance of knowledge of the social sciences to the issue of the sustainability of streams and rivers.
7. Discuss the relevance of knowledge of sociology to the roles that civil engineers play in city planning.
8. Discuss the relevance of knowledge of psychology to the roles that civil engineers play in developing evacuation plans for residents who face an approaching major hurricane.
9. Discuss the relevance of knowledge of economics to the roles that civil engineers play in renovating blighted neighborhoods.
10. Discuss the relevance of knowledge of political science to the roles that civil engineers play in controlling pollution in suburban environments.
11. Discuss the relevance of knowledge of geography and the spatial distribution of land use to the civil engineer's role in controlling urban sprawl.
12. Discuss the relevance of knowledge of social psychology to the civil engineer's role in controlling urban sprawl.
13. Discuss the relevance of knowledge of personality, which is a subdiscipline of psychology, to possible solutions to traffic control and road rage.
14. Discuss the relevance of knowledge of cognitive development on the ethical maturity of engineers.
15. Discuss the relevance of knowledge of organizational conflict to civil engineering businesses.
16. Discuss the relevance of knowledge of motivation theories to productivity in civil engineering businesses.
17. Examine Alexis de Tocqueville's *Democracy in America* and its views on American culture and independence. How do these characteristics affect engineering projects such as housing developments or mass transportation?
18. For each social science discipline discussed in this chapter (psychology, sociology, economics, and political science) propose an idea for new knowledge.

19. Select one of the following problems and analyze it from the perspective of social science disciplines: (a) urban sprawl, (b) parking and traffic in city centers, (c) stream restoration, (d) tidal (wave) power as an energy source, (e) airline safety.

3.11 TEAM PROJECTS

1. Currently, the automobile is the primary means of individual travel. Yet, the automobile is a very inefficient travel mode, at least in terms of the use of nonrenewable natural resources. Propose a personal travel mode of the future and identify how knowledge of the social sciences could make this an acceptable travel mode.

2. In the United States, the single-family home is the housing goal of most people. In other societies where land is more costly, multifamily housing is the norm. Discuss these alternatives from cultural, economic, sociological, psychologic, and political perspectives.

3. Engineers Without Borders performs engineering service activities in underdeveloped areas. Discuss in detail how the projects could benefit from having social scientists as part of the teams.

REFERENCES

Ashforth, B. E., and Saks, A. M. 2002. Feeling your way: Emotional and organization entry. In *Emotions in the workplace*, ed. R. G. Lord, R. J. Klimoski, and R. Kanfer, chap. 10. San Francisco: Jossey-Bass.

Wooden, J., and Sharman, B. 1975. *The Wooden–Sharman method*. New York: Macmillan Publishers.

4 Experimentation

CHAPTER OBJECTIVES
- Discuss the steps of the experimental process
- Outline the analysis of experimental data
- Identify reasons why experimentation is needed
- Discuss criteria used to evaluate the success of an experiment

---

4.1 INTRODUCTION

Consider the following general questions:

- At what speed does an ant move when it is foraging for food? How about when it feels threatened?
- How fast does water evaporate from a dish placed in the sun? In the shade? On a humid day? On a dry day?
- From a batch of balloons manufactured with the same material, how big (volume or maximum width) will they be just before they burst?

How could we get answers to these general questions? Now consider the following questions that may be of interest to civil engineers:

- How much gasoline could be saved in the United States if all traffic lights were equipped with sensors to minimize the wait time at intersections?
- How effective are storm water management ponds in reducing pollution from urban runoff?
- How would the inclusion of small pieces of used car and truck tires as aggregate into concrete roadways affect the likelihood of potholes developing in the roadway?

Again, how could we get answers to these questions? How reliable would the answers be? Would the answers be sufficiently accurate such that we would feel comfortable in using the answers to set design standards?

Experimentation allows us to get answers to the six questions posed above. If experiments were conducted for any of these questions, we would want accurate and timely answers. Following an orderly, established procedure for conducting the experiments would likely lead to the most accurate answers. Understanding the process of experimentation is necessary to get accurate and prompt answers.

Experimental results are necessary for the civil engineering profession to provide accurate designs and solutions for the myriad problems of today and those that will arise in the future as technology advances. As new technologies become available, side effects can create problems with which the design engineer must contend. As new materials are introduced, engineers will need controlled experiments to understand their characteristics, especially under extreme conditions. The effects of accumulated environmental pollution and alternatives for mitigating the effects will require experiments, with the control occurring either in the laboratory or through the application of statistical methods. These are just a few of the issues that make knowledge of the design, conduct, and use of the experimental process important to the practicing civil engineer.

4.2   VALUE ISSUES IN CONDUCTING EXPERIMENTS

Engineering design standards and public policies can depend on the outcomes of experiments. Therefore, those who conduct experiments must be honest, industrious, and recognize the responsibilities associated with conducting experiments and reporting the results. The following are a few of the values that are relevant to experimentation:

- *Knowledge*: Understanding gained through study or experience.
- *Accountability*: Answerable in meeting responsibilities, with some sense of liability for proper conduct.
- *Care*: Conscientiousness and attentiveness to detail.
- *Safety*: Free of danger, risk, or injury to persons or the environment.
- *Perseverence*: Steadfastness; persistence.

A goal of experimentation is to gain new knowledge, but it is important that care and safety be taken so that the results are accurate and reliable. The experimenter is accountable for all aspects of the work, and as problems often arise in experimental work, the experimenter must persevere.

4.3   ELEMENTS OF EXPERIMENTATION

To fully understand the importance of experimentation to the civil engineering profession, each of the following elements needs to be mastered:

- *Knowledge*: To understand the procedure of experimental analysis and synthesis.
- *Comprehension*: To recognize the value of experimentation and to have the ability to explain the procedure.
- *Application*: To be capable of designing and conducting an experiment.
- *Analysis*: To understand how to analyze experimental data and to evaluate the accuracy of the findings.
- *Synthesis*: To have the knowledge to create an experiment that will solve a civil engineering problem.
- *Evaluation*: To evaluate the extent to which an experiment and the results met the original goal.

Of course, mastering each of these elements takes years of experience; however, the fundamental elements can be easily understood.

4.4 THE SCIENTIFIC METHOD

Most civil engineers are familiar with the term *scientific method*, which is a procedure of observation, hypothesis, testing, and analysis. The names of the steps vary with the presentation, but the scientific method is generally an attempt to reflect a systematic, proven method of discovery. It has its modern origin in Francis Bacon's (1561–1626) work, and is widely reported to have significantly contributed to many of the important discoveries of the seventeenth through nineteenth centuries. For our purposes here, we will see that the procedure of engineering experimentation has many similarities to the steps of the scientific method.

Briefly stated, the steps of the scientific method are

- *Observation*: To recognize a problem.
- *Hypothesis*: To formulate a testable statement that will lead to a decision about the problem.
- *Testing*: To design and conduct the experiments needed to prove or disprove the hypothesis and collect relevant data.
- *Analysis*: To analyze the experimental data, develop general conclusions related to the problem statement, and communicate the results to interested parties.

While this process may appear simple, even the work of seasoned scientists may fail if the wrong problem is identified, the hypothesis is not stated in a way that will lead to a solution, or the experiment is not conducted properly. One problem with this four-step process is that the few number of steps may make it difficult to understand exactly what each step encompasses.

4.5 CONDUCTING AN ENGINEERING EXPERIMENT: PROCEDURE

Numerous representations of the procedure for designing and conducting an engineering experiment have been proposed. The alternatives vary in the number of steps, some with as few as four steps and others with as many as a dozen. An eight-step procedure is presented here. It begins with the identification of the problem and ends with the dissemination of the results. The procedure is valid for experimental analyses of both relatively simple problems and more complex problems that have global implications. The following eight-step procedure applies to laboratory experiments, field studies, and computer or virtual experiments:

1. State the problem.
2. State the goal and objectives.
3. Design the experiment(s).
4. Prepare the experiment.
5. Collect the data.
6. Analyze the experimental data.
7. Identify implications of experimental analyses.
8. Report on the study.

CASE STUDY

Mentoring is a very important part of discovery and experimentation. A solution to a problem cannot be found if the proper experiment is not developed. Knowledge of good experimental procedure is often the result of good mentoring. For example, James Watson and Francis Crick are credited with discovering the structure of DNA, but they acknowledged the importance of their mentors. Similarly, Michael Faraday (1791–1867), who is known for his study of magnetism and electrical current, apprenticed under Humphrey Davy (1728–1829), who taught Faraday much of what enabled him to conduct the experiments that led to his discovery.

4.5.1 STATE THE PROBLEM

In many cases, problems are acknowledged only long after they should have been recognized. Ideally, they would be anticipated before they become major issues. For example, the indiscriminant dumping of hazardous wastes created problem sites, such as Love Canal, that were recognized as problems only after extreme health problems caused people to seek the cause. It is not uncommon for people to argue that the problem exists even though they do not have credible evidence to support their hypothesis. The connection between smoking and cancer was recognized a decade before medical evidence was provided. On the other hand, people will argue that the problem does not exist even though they lack credible support. For example, in spite of scientific evidence, some people do not believe that we are experiencing global warming. The actual effect of a problem is not generally agreed upon unless incontrovertible evidence is available. Experiments are the means of gathering the data needed to make a decision, and it is often the engineer's responsibility to lead the effort to obtain credible and accurate evidence.

The way that the problem is stated is very important. It must be specific, accurate, and directly relevant to the desired outcome. An inaccurately stated problem could lead to a poor experimental design that will not lead to a credible result. An accurately stated problem will lead to logical statements of the goal and objectives of the experimental analysis.

4.5.2 STATE THE GOAL AND OBJECTIVES

A single, general, but precise statement of the problem leads directly to the goal of the experiment. By meeting the stated goal, the problem should be solved. The statement of the goal should be followed by a set of research objectives, which are very specific statements that directly relate to the research goal. Note that the statement of the goal is very general, while the objectives are specific statements. The objectives may be expressed either as hypotheses that need to be tested or as effects that will be quantified. By focusing on and providing results for each objective, the research goal will be achieved and the solution to the research problem understood.

Example

Consider the following problem. Pollution from residential areas can be very damaging to the environment. Streams that receive urban runoff are deteriorating in a number of ways. The aquatic life of a stream is especially vulnerable to the storm runoff and the pollutants it transports. One specific source of detrimental pollutants is the grit that erodes from rooftop shingles and its associated chemicals. This is a problem that needs experimental evidence to provide a solution. The experiment could be conducted in a laboratory or from actual rooftops. Based on this background, the following problem statement could be formulated:

Problem: Many pollutants that damage the environment originate in rooftop runoff from rainfall.

This is a general statement of the problem. A more specific statement, such as the following, could also be appropriate:

Problem: The grit that originates from rooftop runoff damages the aquatic habitat of receiving streams.

Either of these problem statements could lead to the following goal:

Goal: To evaluate the effectiveness of cisterns to reduce pollution loads in runoff from rooftops.

Note the generality of this goal, as it does not specifically address (1) the meaning of effectiveness, (2) the characteristics of cisterns that would be important, (3) the amount of pollution reduction, (4) which pollutants need investigation, or (5) the criterion used to measure damage to the habitat. These specific issues are appropriate for the objectives.

It would be possible to develop many research objectives that would help solve the stated problem. The following is a possible objective:

Objective 1: Assess the effects of rainfall intensity (mm/h) and duration (h) on the reduction rate of rooftop grit (lb/ft2/yr or N/m2/yr).

This objective identifies a factor important to effectiveness, as rainfall characteristics differ in Miami, Minneapolis, and Menlo Park, California. Certain intensities would have the power to move the grit. A second objective is:

Objective 2: Evaluate the effects of cistern characteristics (e.g., volume, orifice location) on grit reduction.

Solutions to these two objectives would be a start in solving the problem, yet be general enough to give engineers direction regardless of their location. Again, other objectives (hypotheses) could be formulated to examine other factors (e.g., age of the shingles).

The general statement of the goal has the advantage that those interested in using cisterns as a control method would have the information needed to complete the design. If the goal specifically stated that the experiment was only for College Park, Maryland, then the research would be of less value to the civil engineering profession and ultimately be of less help in solving a broad problem faced throughout the United States or the world.

4.5.3 Design the Experiment(s)

For each research objective, an experiment must be proposed. For each experiment, the following factors must be incorporated into the design:

- Identify all relevant variables that will be varied in the experiments.
- Estimate the required sample size needed to achieve a level of assurance that the results are valid. The sample size is very often controlled by resources or constraints, e.g., lack of time or money.
- Decide on the instrumentation required for measuring the variables.
- Identify the methods to be used in the analysis of the measured data.
- Identify variables that will be held constant throughout the experiments; these are sometimes referred to as environmental factors.

It is generally difficult to specify the exact sample size required because it depends on the variation that will be found in the data. Sample sizes can usually be approximated based on variations found in past experiments or from statistical theory. The sample sizes may need to be adjusted after data collection has begun and some data are collected.

The methods of analysis depend on the statements of the objectives. Some analyses may require a method of parameter fitting such as least squares, while other analyses may only require a statistical test of the hypothesis, such as the analysis of variance on means. Criteria that indicate bias, precision, and accuracy may be needed to provide quantitative support for the experimental results. For example, the correlation coefficient is widely used to indicate accuracy for linear regression analyses. For nonlinear analyses the average error, i.e., the bias, is frequently used as a second goodness-of-fit index.

Considerable thought should be given to identifying the details of an experimental design. An inadequate sample size, failure to adequately control the environmental factors, collecting data over a small range of the experimental variable, or selecting an inappropriate statistical method can lead to incorrect answers to the problem. For example, the use of an unsuitable hypothesis test may suggest an insignificant effect.

CASE STUDY

The rise of experimentation started in the early seventeenth century. Prior to this, much of science was conducted through rational thought. Economic factors often motivated experimentation and innovation in the fifteenth and sixteenth centuries. The need to improve economic competitiveness influenced artisans who lacked a formal education to find ways to improve production processes. Ultimately, it was Francis Bacon (1561–1626) and René Descartes (1596–1650) who formalized the steps of systematic experimentation. The experimental procedure formalized during that period made the industrial revolution possible. Many of the technological advances of today are the result of the process proposed by Bacon and Descartes.

When the true finding of a significant effect has been identified, the correct solution is more likely to be found. Erroneous designs may result from a lack of knowledge or experience, or from not taking the time to properly design the experiment.

4.5.4 PREPARE THE EXPERIMENT

Once the experiment has been designed, preparation will be needed. Instruments that will be used for measurement must be obtained and calibrated. In field experiments, such as in the installation of a weir, the equipment must be installed and calibrated. If analyses are to be made using computer programs, commercially available programs must be identified or new programs must be written. The programs should be fully tested prior to use.

4.5.5 COLLECT THE DATA

One advantage of laboratory experiments is that better control of both experimental and environmental variables is often possible. This reduces some of the sampling variation, thus increasing the accuracy of the results. Laboratory experiments may also allow for control of the intercorrelation among variables. With field studies, conditions such as the weather cannot be controlled; it is more difficult to estimate the time frame required to complete the experiment.

As data are collected, analyses of the data should begin immediately in order to assess whether or not changes will be needed in the experimental design or the proposed data collection scheme. This provides feedback that may even necessitate revision of the objectives as well as the experimental design. Analysis of initial data may suggest that the designed experiment will not be able to meet the objectives, in which case an increased amount of data will be needed or other variables included in the experimental design.

4.5.6 ANALYZE THE EXPERIMENTAL DATA

Once the database has been assembled, each of the analyses indicated in the experimental design and suggested in the research objectives should be made. Initially, this may involve graphical analyses of the data, such as boxplots, x–y graphs, or histograms.

CASE STUDY

During the sixteenth and seventeenth centuries, the way that experimentation was conducted underwent a fundamental change. Experimentation took a more quantitative nature, which necessitated the need for new ways of measurement. Torricelli (1608–1647), an Italian scientist, invented the mercury barometer; Galileo (1564–1642) invented the thermometer; and Gunter (1581–1626) invented the slide rule. Inventions such as these enable precise data collection and quantitative effects to be identified.

These graphical compilations would likely be accompanied by computations of statistical characteristics such as means and standard deviations. If independent data sets are to be used for statistical validation, then such analyses must be made.

4.5.7 Identify Implications of Experimental Analyses

Once the statistical analyses have been completed, the results must be interpreted. The first and most important analysis would be to assess the rationality of the results. If they defy expectations, then the computations should be checked to ensure that they were done correctly. This would include checking the validity of the computer programs used to assist in the computations. The inability to control environmental factors could contribute to irrational results and may necessitate additional testing. Unexpected results should not be used to make decisions until the cause is determined. After all, the unexpected results may be correct, while the initial expectations were wrong.

The data analyses may affirm or refute the existence of the effect or trend specified in one of the research objectives. Such results have practical meaning, and the implications of the findings for others who have an interest in the results should be identified. Very often, the implications of the results are as important as the numbers themselves.

4.5.8 Report on the Study

It can be argued that experimental work that is not disseminated, such that the advance of knowledge is not shared, is of minimal value. Therefore, good communication skills are necessary for the report of the work to be clearly summarized. The reader of the report must understand what work was done, any limitations of the experiment, and the implications of the results. Appendix A includes guidelines for writing reports.

4.6 APPLICATION OF THE EXPERIMENTAL PROCEDURE

The following brief application illustrates the steps of the experimental procedure.

4.6.1 Problem Statement

Hydrologic design often requires estimates of the travel time of runoff, which is dependent on the velocity of the flow. The velocity of runoff depends on the slope and roughness of the surface and the depth of flow, which can be dependent on the rainfall intensity. If the watershed is urbanized, the flow path of the runoff might be partly over a grassed lawn, then into a gutter, and then through an inlet into a pipe storm sewer. Each of these paths will have different characteristics, with different velocities.

The velocity of flood runoff is especially important because it can cause a number of problems, including

1. Scour and transport of pollutants from roadways, parking lots, and rooftops
2. Scour of channel sides and bottom
3. The transport of trash
4. The amassing of runoff at inlets, which causes local flooding

The velocity of flow is often estimated using Manning's equation, which involves a runoff coefficient, values of which are generally taken from a table. However, they are not very accurate. More accurate estimates would yield better estimates of runoff velocities, which could lead to better estimates of scour and flooding.

4.6.2   GOAL AND OBJECTIVES

The goal of the study is to characterize the surface resistance to flood runoff on selected surfaces. The objectives of the study are:

1. Estimate the roughness coefficients for land surfaces of asphalt and concrete gutters.
2. Compare the experimental values with values commonly available in tables.
3. Evaluate the site-to-site variation of the roughness coefficient.

4.6.3   EXPERIMENTAL DESIGN

Tabled values of the roughness coefficient indicate a coefficient of variation (standard deviation divided by the mean) of 20%. Thus, the standard deviation (σ) is 20% of the mean, and normal theory would suggest a sample size (n) of

$$n = \left(\frac{z\sigma}{e} \right)^2$$

where z is the standard normal deviate for a 5% level of significance, and e is the tolerable error in the mean, where the tabled value is used as an estimate for the mean. For means of 0.016 for rough concrete and 0.012 for asphalt, a tolerable error of 0.02 and a z value of 1.96 yield sample sizes of 10 for concrete and 6 for asphalt. The sites should be selected at random.

Because the slope of the gutter and the amount of runoff, or rainfall, influences the velocity of flow, it would be necessary to collect the same number of samples for each slope range (e.g., shallow, moderate). To reflect the amount of runoff, the depth of flow in the gutter should be measured, which can then be used to compute the hydraulic radius. The equipment needed to collect the data includes a flow measurement device and a meter stick for length measurements (e.g., flow depth and the rise and run to compute the slope).

4.6.4   DATA COLLECTION

The data collected are shown in Table 4.1 for the four conditions: (1) concrete, low slopes; (2) concrete, moderate slopes; (3) asphalt, low slopes; (4) asphalt, moderate slopes.

4.6.5   ANALYSIS OF DATA

The samples show some variation, which is referred to as sampling variation. For example, the ten measurements of roughness on low-sloped concrete pavements

TABLE 4.1

Roughness Coefficients for Asphalt and Concrete Surfaces at Low and Medium Slopes

Concrete		Asphalt	
Low	Med	Low	Med
0.011	0.010	0.010	0.009
0.012	0.010	0.010	0.010
0.013	0.012	0.010	0.010
0.013	0.012	0.010	0.012
0.014	0.013	0.012	0.012
0.015	0.014	0.012	0.012
0.016	0.014	0.014	0.013
0.017	0.017	0.014	0.013
0.017	0.018	0.014	0.014
0.019	0.018	0.014	0.015

ranged from 0.011 to 0.019 with a mean of 0.015. The standard deviation is 0.003. To decide whether or not the sample means differ from the assumed population means, which are the values taken from tables of roughness coefficients, the one-sample t test can be used. The means for the asphalt samples are the same as the population value, i.e., 0.12, so we can assume that the samples are representative of the population. The means for the concrete surfaces are both less than the population mean. The computed t statistic for the moderately sloped concrete surfaces is

$$t = \frac{\bar{x} - \mu}{s/\sqrt{n}} = \frac{0.0138 - 0.016}{0.003/\sqrt{10}} = -2.31$$

With 9 degrees of freedom, the critical t value for a two-sided test and a 5% level of significance is ±2.26. Since the sample value of −2.31 falls outside the range from −2.26 to +2.26, the null hypothesis of equality of means is rejected. This indicates that the sample values do not reflect the generally accepted value of 0.016.

To examine the site-to-site variation, a two-sample t test can be conducted between the low and moderately sloped gutters. Since the sample means for the two asphalt sites are the same, the t statistic would be zero and the difference is obviously not significant; i.e., the roughness of asphalt is not dependent on slope. The pooled standard deviation (s_p) for the concrete measurements is

$$s_p = \left[\frac{(10-1)(0.00254)^2 + (10-1)(0.00301)^2}{10+10-2} \right]^{0.5} = 0.002785$$

Therefore, the t statistic is

$$t = \frac{0.0147 - 0.0138}{0.002785(0.1 + 0.1)^{0.5}} = 0.723$$

For 18 degrees of freedom, the critical t value is ± 2.101. Therefore, the null hypothesis of equality is accepted.

These statistical calculations indicate a dilemma. The moderately sloped roughness was found to be different from the tabled value, yet the low-sloped values are not different from the tabled value. However, the two-sample t test indicates the two sample values are equal. The solution to this problem of logic is that the use of a 5% level of significance, which is very common in engineering, is somewhat arbitrary. If a 1% level had been used in this case, then the dilemma disappears. Based on this, the evidence does not seem strong enough to reject the idea that the tabled values of roughness are acceptable.

4.6.6 IMPLICATIONS OF THE ANALYSIS

This experiment was simple and narrowly focused on a limited type of surface under a narrow range of conditions. Therefore, the implications of the study would be limited, such as

- The procedure seems applicable for measuring the roughness of shallow- or medium-sloped gutters.
- The small samples seem adequate.
- The tabled values of asphalt and concrete roughness are correct.
- The variation, as measured by the coefficients of variation, is large.

Based on these results, it seems that the procedure could be applied to evaluate the roughness of other surfaces.

4.7 FACTUALITY, RATIONALITY, AND ACTUALITY

Having completed an experiment and summarized the results, often using statistical methods, evaluating the results is critically important. Three criteria will be used:

- *Factuality*: The results should agree with all measured data and observations.
- *Rationality*: The results should conform to the understanding of the physical system.
- *Actuality*: The results should increase knowledge by overcoming misperceptions and allowing a reorganization of beliefs.

The database obtained as part of the experimental design is unlikely to be the only available data or observations. The experimental results should be evaluated using other data and observations. This is sometimes referred to as independent verification like the experimental analysis, statistical methods are widely used in testing the experimental results with the independent data. It is generally believed that the goodness-of-fit results to this independent data will suggest a lower level of accuracy because of differences in conditions. Those involved in the research must decide what loss of accuracy is acceptable before the verification casts doubt on the experimental results.

Rationality is the second evaluation criterion. This criterion is less quantitative, as it is not based on statistical measures. Instead, the evaluation uses the processes of the physical system. The degree to which the experimental results conforms to expectations is the idea underlying the evaluation. For example, do evaporation rates increase with temperature? Does the beam deflection increase with increasing load? If the experimental results show an effect that does not conform to physical expectations, then the experimental results are not verified.

As an evaluation criterion, actuality goes beyond factuality and rationality. The experimental analysis was conducted to advance knowledge. This often involves overcoming misperceptions so that one's knowledge can be reorganized, with the new knowledge providing a more rational understanding of the processes. Actuality is intended to give breadth to knowledge, just as the data and observations of the factual assessment extend the analysis quantitatively.

Other evaluation criteria can be used, such as implications to cost, natural resource, use, public welfare, and ease of application. Each of these evaluation criteria will need to be justified as being a legitimate metric.

4.8   DISCUSSION QUESTIONS

1. Identify a problem in the natural world, such as the velocity at which ants forage, and design an experiment to evaluate the problem.
2. Identify a problem from the arts, such as the force used in sculpturing, and design an experiment to evaluate the problem.
3. Perseverance was identified as a value relevant in experimentation. Discuss its importance and necessity.
4. Accountability was identified as a value relevant in experimentation. Explain how accountability is a connection between ethics and experimentation.
5. Obtain dictionary definitions of the words *analysis* and *synthesis*. Discuss their relevance to the development and application of a civil engineering design method.
6. Develop a two-hundred-word bibliographical summary of Francis Bacon, with specific reference to his contribution to experimentation.
7. Outline the use of the eight-step process of Section 4.5 in conducting a field experiment intended to compare the effects of land development (i.e., the construction of a residential community) on the quality of a small stream that receives storm runoff, before and after the development.
8. Identify the equipment necessary to determine the compression strength of a new material. Establish the procedure that you would use to make the necessary tests.

9. Identify the equipment necessary to measure the dissolved oxygen content of a river. What procedure should be followed in collecting and analyzing the data?

10. Assume that traffic rates and decisions of drivers (turn left, turn right, go straight ahead) were needed at a four-way intersection. Design an experiment to study this issue.

11. Assume that estimates of evaporation rates from lakes and ponds used to provide irrigation water are needed. Design an experiment to determine the rates as a function of important factors.

12. Assume estimates of corrosion rates of a particular type of steel were needed. This would be a long-term study with the specimens exposed to the natural environment. Design an experiment that would provide estimates of corrosion rates as a function of time and environmental (weather) conditions.

13. Outline a report that would summarize the experiment of discussion question 10.

14. Outline an oral presentation that would summarize the experiment of discussion question 12 for a technical audience.

15. Outline an oral presentation that would summarize the experiment of discussion question 11 for a nontechnical audience.

16. Identify statistics that can be used to measure the bias, precision, and accuracy of research results.

17. Discuss the construction of a histogram for presenting experimental data. Develop rules that will ensure a histogram will clearly indicate the characteristics of the data.

18. If a linear model $y = a + bx$ is fitted to experimental data, discuss the meaning of the two coefficients a and b.

19. Discuss the use of hypothesis tests in the interpretation of experimental results.

20. Identify criteria that can be used to evaluate the success of an experimental design. These can be statistical criteria or metrics based on the physical processes involved in the experiment.

21. Most statistical methods are based on assumptions, e.g., large sample size, normal distribution. With most engineering data, these assumptions are not met. If this is the case, can statistical criteria be used as evaluation metrics when assessing the success of an experimental design?

4.9 GROUP ACTIVITIES

1. Design and conduct an experiment to demonstrate Archimedes' principle. The experiment must be designed as a model of a pontoon bridge. Then show the use of Archimedes' principle to design a pontoon bridge to support a 300 kN tank.

2. Design and conduct an experiment to show the effect of the pulling force needed to detach tape from a table as a function of the tape width and the angle at which the tape is pulled. Use three types of tape (e.g., masking, packing, and adhesive). Use a spring scale to measure the required force to

separate the tape from a piece of wood (or tabletop). The angle at which the tape is pulled is measured from the horizontal.

3. Design and conduct an experiment to show the modulus of elasticity. Use a metal meter stick as a bridge (position the meter stick between supports, such as blocks of wood) and bronze masses as the load to cause the meter stick to deflect.

5 Sustainability

CHAPTER OBJECTIVES
- Discuss values and ethical issues relevant to sustainability
- Discuss the creation of new knowledge of sustainability
- Present criteria for evaluating sustainable development

5.1 INTRODUCTION

Worldwide consumption of petroleum products is increasing. Developing countries need petroleum to fuel this economic growth. However, the supply is limited, and political and economic pressures at international levels often limit the extraction amounts. Furthermore, the efficiency of the use of petroleum products has not been a primary concern, as shown by the sheer number of gas guzzling cars produced for U.S. consumption. In summary, demand is up, the production is often constrained, and the supply is limited. Should we be concerned for future generations? In the future, will shortages of petroleum drastically constrain economic growth, maybe even cause severe consequences to human welfare and health? Can we constrain our current consumption to ensure that future generations will have a reasonable supply? Should we now place greater economic resources on developing alternative energy forms? Should we even worry about future generations? These are questions that are relevant to sustainability and the civil engineering profession.

Sustainability is a contemporary issue that affects the future of engineering, particularly civil engineering. Engineers can certainly influence the extent to which our communities sustain air, soil, water, and energy resources. Then, what is meant by sustainability? How will it influence the career paths of current undergraduate civil engineering students? What are value issues inherent to sustainability? Why is it considered an ethical issue?

Petroleum is not the only natural resource about which questions that are important to society are being asked. Air, soil, flora and fauna, and surface and groundwater resources are imperiled. Recent concerns about polar bear habitat are one example. At times in the past, game fish were abundant in most streams and rivers. Now, many streams and rivers support very little aquatic life, which is a sign that development has had unintended consequences. The dead streams can no longer sustain life, which is an indication that development failed at ecological sustainability. These examples clearly illustrate that sustainability must be viewed from a broad perspective.

5.2   SUSTAINABILITY: DEFINITION

The American Society of Civil Engineers (2008, 157) defines sustainable development as "the challenge of meeting human needs for natural resources, industrial products, energy, food, transportation, shelter, and effective waste management while conserving and protecting environmental quality and the natural resource base essential for future development." In 1996, the ASCE modified its code of ethics to emphasize the importance of sustainable development and challenged civil engineers to comply with the principles of sustainable development in every aspect of their professional duties. As the ASCE Code of Ethics recognizes many responsibilities, this of course requires in some way balancing sustainability with other competing professional responsibilities. The sustainable use of natural resources may place certain aspects of public safety at risk. Since public safety is also a principal element of the ASCE code, the engineer will need to have sound judgment to fulfill these competing responsibilities.

Four types of sustainability are

- *Energy resource sustainability*: Natural resources such as petroleum, timber, and coal are being rapidly consumed. They cannot be replaced in a time frame that will provide ample supplies for immediate generations. It is vital that we minimize the use of these energy sources, and research, develop, and use alternate renewable energy forms.
- *Ecological sustainability*: We must maintain the health of ecological systems to ensure sustainable populations and the diversity of flora and fauna. The choices that humans have made have damaged some aspects of the health of our planet. Many ecosystems have been stressed by nearby urban and suburban development and pollution. For example, many dams have been constructed for multipurpose benefits (power, flood control, recreation), but they have limited the passage of both fish to spawning grounds and fertilizing silt to the river deltas. Some dams are now being removed in order to restore natural passageways and promote ecological sustainability. Engineers must recognize that their actions can have unplanned or unrecognizable consequences. Thus, engineering planners must critically assess each project for potential short- and long-term problems.
- *Soil sustainability*: Soil is a natural resource that must be protected. Soil is the foundation material for many civil engineering projects, such as earthen dams and roadways. Its properties are important in storm water best management practices. In a very short period of time, soil fertility can be significantly reduced by human actions. Soil has many uses, including agricultural production, but misuse can leave the soil much less productive for future generations. Many ecosystems depend on soil, such as forest, wetlands, and grasslands. Soil sustainability must be viewed in terms of both topsoil losses due to erosion and transport away from the site and negative changes to its chemical composition. Solving these problems requires engineering research and recognition of the importance of soil sustainability.

CASE STUDY: SOIL SUSTAINABILITY

Traditionally, soil sustainability has implied the maintenance of soil resources such that agricultural productivity was not constrained. The view of soil as a resource has changed and is continuing to change with time. Specifically, the importance of soil in controlling water quality and enhancing biodiversity is no longer judged solely on its rock and mineral content. Its nutrient content, the level of organic matter, and its ability to sustain both are now accepted criteria for measuring soil sustainability.

Soil sustainability is relevant to civil engineering. Soil serves as the foundation for structures and roadways. Soil masses in bioretention facilities are the basis for water quality control. Soil composition is important as a building material, such as in earthen dams.

- *Environmental sustainability*: The impacts of urban and suburban development have been well documented, most notably increased flooding, reduced groundwater recharge, increased erosion of streams and rivers, and degrading of water quality. Pollution was so bad at one time that the surface of a river in Ohio caught on fire. While rivers and streams are ecologically important, they are also important for transporting wastewater and providing clean water. Therefore, the quality of rivers and streams must be sustained to ensure that they can mitigate wastes and provide drinking water effectively and at a reasonable cost.

5.3 WHY SUSTAINABILITY IS IMPORTANT

Brundtland (1987) provided an alternative definition of sustainable development that emphasizes the balance of the needs and aspirations of the present with the needs of future generations. This definition is more narrowly focused than the ASCE definition, as Brundtland placed a direct emphasis on the needs of future generations. This illustrates the breadth at which sustainability must be viewed. The current generation has a legitimate right to use our natural resources, but at the same time, it has a responsibility to consider the needs of future generations when using natural resources.

Sustainability is closely associated with the environment, just as engineers are linked to the environment. The choices that engineers make in project design can affect the health of natural systems, as well as the use of energy resources. For example, infrastructure projects are often materials-intensive, many of which fall in the nonrenewable class; e.g., asphalt for roadways is petroleum-based. We must conserve natural resources as well as minimize damage to existing natural processes. Through design choices, engineers have a central role in influencing environmental quality both for the present and for availability to future generations.

CASE STUDY

Extensive farming without crop rotation coupled with severe drought created brutal dust storms in the Great Plains during the 1930s. The storms lifted topsoil and carried it as far as New England and the Atlantic Ocean. American farmers were forced to relocate as their homes were destroyed and their land was subject to foreclosure. Erosion preventative farming was implemented in the mid-1900s. Today, the Great Plains is a sustainable agricultural producer, with considerable tons of agricultural products exported. This is an example of the benefits of sustainability.

Today, sustainability is an issue that garners increased attention and controversy. As a civil engineer, it is essential to be able to

- Understand the values and ethics that underlie sustainability
- Comprehend sustainable development
- Understand the impact of engineering works on natural systems
- Apply the ethics of sustainability to infrastructure development
- Evaluate the sustainability of complex systems

Developing these abilities and understandings should begin as part of the undergraduate civil engineering program and continue as part of lifelong learning.

5.4 SUSTAINABILITY AND HUMAN VALUES

Sustainability has a technical basis and can be discussed in terms of natural resources, the physical processes, and energy. However, sustainability is also intertwined with human values. Recognizing the ways that sustainability can be accomplished without understanding its value basis is inadequate. Engineers who see sustainability from a value perspective will be more able to strive to incorporate all of the principles of sustainability in their work.

The following are a few of the human values that are relevant to sustainability:

- *Fairness*: Impartial; just to all parties.
- *Duty*: A scrupulous sense of responsibility in carrying out tasks.
- *Knowledge*: Understanding gained through study or experience.
- *Efficiency*: The quality of producing effectively with a minimum waste of resources.
- *Welfare*: The general well-being of the public and the environment.
- *Accountability*: Answerable for duties and obligations, with some sense of liability for proper conduct.

Each of these values applies to the individual engineer and the engineering profession. In turn, society is responsible for providing the atmosphere and resources so that engineers can fulfill their value responsibilities.

Sustainability requires fairness among the current generation and generations of the future. It requires fairness between competing responsibilities, such as the use of nonrenewable natural resources and providing public amenities. Some engineering projects may need to ensure public safety at the expense of sustainability when the overall good of the project is recognized.

Engineers have a duty to be knowledgeable about sustainability and the ways to incorporate it into design projects. Engineers cannot automatically place sustainability as a distant second to other responsibilities cited in the code of ethics. While sustainability is a relatively new idea, this does not mean that it is a less important duty.

Sustainability appears to be less well defined than value responsibilities like public health or safety. Therefore, gaining knowledge of sustainability will require greater effort than knowledge of the more traditional value responsibilities. A precise, all-inclusive definition of sustainability is not possible, so engineers may need to exert greater effort to understand the full breadth of sustainability.

When using nonrenewable natural resources, efficiency may be the most important value with respect to sustainability. To achieve sustainability, the engineer needs to be knowledgeable about the renewable resources options. This is part of the commitment to efficiency. Note that the value responsibilities are not independent of one another. Efficiency is closely related to knowledge and duty.

Human welfare is one value responsibility that is traditionally part of the engineering codes of ethics. Sustainability requires the interpretation of value welfare to apply to the environment as well as to human well-being. Also, the welfare of future generations, as suggested by Brundtland (1987), is part of sustainability. Welfare must be viewed from this broader perspective, not the narrow perspective of the welfare of humans of this generation.

Accountability is included as a value relevant to sustainability because engineers are responsible for all values identified in the code of ethics. Engineers have an ethical obligation to fully address sustainability. Proper conduct in design includes being accountable to the efficient use of nonrenewable resources.

CASE STUDY

The Nile River system has been a sustainable agricultural system for thousands of years, with yearly flooding that delivers fertile silt to areas where it is needed to sustain farming. The Aswan Dam has interrupted this process, and farmers have experienced declining productivity. As a result, farmers have resorted to synthetic fertilizers, which have polluted water supplies, degraded the groundwater quality, and harmed the fishing industry. Many worry that the Three Gorges Dam on the Yangtze River, China, will also cause animal extinction, erosion, and drown many historical artifacts. These are examples where better planning for sustainability was needed.

5.5   ETHICS OF SUSTAINABLE DEVELOPMENT

The fundamental canons of the ASCE Code of Ethics (1996) include the statement: "Engineers shall ... strive to comply with the principles of sustainable development in the performance of their professional duties" and "recognize that the lives, safety, health, and welfare of the general public are dependent upon engineering judgments, decisions, and practices." It is clear that engineers cannot ignore issues related to sustainability and have a responsibility to create and maintain sustainable systems by reducing the environmental impact of projects and designs.

Engineers have multiple ethical responsibilities, as indicated in the ASCE Code of Ethics. While they have principal responsibilities to the public in terms of health, safety, and welfare, engineers also have important responsibilities to the profession, their employer and clients of their employer, and themselves. The engineer's ethical responsibilities also extend globally, and the impact of their designs on the world population of people, as well as animals and plants, must be considered. Ensuring that they simultaneously meet all of these ethical responsibilities is not easy. Sustainability is actually inherent to each of these responsibilities. Sustaining natural resources and the environment benefits society in many ways, including aesthetics, public health, and preservation for future use. The image of the profession is enhanced, and members are more highly respected if the profession is recognized as being "green." While society can establish public policies to promote sustainability and the profession can promote it, sustainable development depends on individual engineers using sustainable practices and employers of engineers providing the encouragement and support of green design.

Engineers must hold themselves to a higher standard than just compliance with laws and regulations. Professionals must take responsibility before public policy demands by asking the right questions and raising relevant issues. It is also important to be conscious of sustainability issues and keep pace with new developments through lifelong learning gained from media coverage and meetings, seminars, and workshops of professional trade associations.

The cumulative effects of individual projects can affect the sustainability of our physical systems, and the design engineer has a responsibility to ensure that the project being designed will not have adverse environmental effects when considering other engineering projects, existing or planned. For example, an engineer who is involved in land development is often required to provide storm water management facilities needed to control storm runoff from the developed site. Even if the project provides adequate control of runoff, the engineer has a responsibility to consider if, when built with other necessary engineering works (e.g., new state roadways, service facilities that will subsequently be considered), the project will significantly increase erosion of the local stream system. By itself, the project will not adversely affect the stream, and even if the other planned projects provide controls, the combined effects will cause damage to the stream. This should then be an ethical concern of the design engineer, as the sustainability of the stream ecology would be in question.

5.6 SUSTAINABILITY AND THE TRANSFER OF TECHNOLOGY

Efficiency was identified as a key value with respect to sustainability. Technology is generally thought of as a means of increasing efficiency, but efficiency may decrease if the technology is beyond the capability of the user. For example, if technology related to agriculture is transferred to another part of the world, where people cannot use the technology properly, then the new technology may hinder the objective of sustainable agriculture. If tractors are used to increase the area of cultivated land to increase food production, the plowed areas may actually cause greater soil losses than with conventional farming practices if left uncovered. The depletion of soil productivity will not be sustained over the long term, so the introduction of the technology is counter to the goal of sustainability. The humanitarian group Engineers Without Borders plans their designs according to the capabilities of the location. The technology must be compatible with local power sources (e.g., charging systems for solar panels), and the material must be locally available for repair. In addition, there will need to be members of the community who are able to maintain and repair the technology.

5.7 CREATING NEW KNOWLEDGE

New knowledge will be needed to achieve a fully sustainable economy and environment. Efficiencies of renewable energy methods are improving, yet certainly do not compare to the rate of increase in energy demand. Other energy generation methods will be needed to achieve success as a sustainable world.

New knowledge with respect to sustainability will be needed in all areas of civil engineering. Zero-energy buildings, which are ones that do not use energy from the power grid, will be important to meet the goal of sustainability. Changes in transportation modes remove use from the one-person-per-car philosophy and will greatly reduce our dependence on fossil fuels. Greater acceptance of the use of gray water could reduce the depletion of groundwater resources. Similarly, the reuse of building materials from buildings being torn down could reduce dependency on raw materials. Innovative thinking by civil engineers can create real solutions for these civil engineering issues. A few specific ideas that may improve our stance on sustainability include:

- *Air quality*: Examine how alternative smart growth land development patterns could reduce greenhouse gas emissions.
- *Materials*: Develop a forest management strategy toward sustainable wood products that accounts for both natural and human deforestation.
- *Groundwater*: Examine how more intense storms that are expected under global climate change scenarios will affect the sustainability of groundwater resources.
- *Soil quality*: Examine how increased use of biofuels will affect soil quality for future food production demand.

CASE STUDY

Trees are a renewable resource, but many forest ecosystems are in danger. Ecologically effective management methods will require new knowledge about the biological characteristics of trees. Reducing the time required for trees to grow will require new knowledge about optimum fertilization, new tissue culture methods, and the effects of soil characteristics on growth rates. Gaining new knowledge that will ensure sustainability of forest resources will require research about both the material properties of new wood products and greater demands on water resources where irrigation is required for optimum growth.

New knowledge in all areas of civil engineering can help us meet global sustainability criteria.

5.8 EVALUATING SUSTAINABLE DEVELOPMENT

Sustainable technologies are designs that reduce the negative impacts on the environment and ensure a sustainable future. Ideas like green roofs and best management practices are such technologies. To evaluate the sustainability of specific technologies, many views must be considered. The following three questions are especially relevant when addressing an issue where sustainability is a concern.

5.8.1 WHAT RESOURCES ARE USED?

Sustainable technologies should minimize the use of nonrenewable natural resources, the energy used for production, and materials used in the design. This can be accomplished by increasing the use of renewable or recycled resources, and recycling or reusing excess materials created during the production process. The design of the product should also be durable, long lasting, and built with replaceable components. Packaging requirements should also be reduced to conserve materials. These principles will maximize the efficiency of resource use.

5.8.2 WHAT IS THE LIFE CYCLE OF THE PROJECT?

The engineer should consider the full life cycle of a project and, if possible, make design changes that will extend the life cycle. Whether a project will last ten or one hundred years, the civil engineer must consider the long-term effects. The need for early replacement or excessive maintenance creates unnecessary demands on resources that could be saved or reused. The biodegradability of the materials used must also be considered to reduce waste. Long-term effects are often unforeseeable, such as the increased rate of birth defects around Love Canal or the disruption of animal migration patterns by petroleum pipelines. Knowledge of historical aspects of technology and engineering can help foresee problems that may arise in the future. Engineers need to think critically about potential long-term effects.

CASE STUDY: SUSTAINING WATER RESOURCES

Hydrologic engineering, which is an important discipline within civil engineering, deals with elements of the hydrologic cycle, which includes processes such as precipitation, infiltration, surface runoff, and evapotranspiration. With respect to water resources, sustainability implies the maintenance of a natural balance between these elements. Since sustaining forests and wetlands is important to many of these hydrologic processes, hydrologic engineers must consider the effect of terrestrial changes on the elements of the hydrologic cycle.

Civil engineers help sustain water resources through the development of best management practices for controlling water quantity and quality, consider the effect of their design on the elements of the hydrologic cycle, help develop public policies that limit damage to the environment, properly operate water resource facilities such as large dams that affect runoff processes in rivers, and properly manage water resources for water supply.

5.8.3 WHAT ARE THE ECOLOGICAL CONSEQUENCES?

Ecological sustainability is a general goal. The civil engineer must consider the impacts of a project or proposed design upon the immediate ecosystem, as well as the global environment. For example, will a project result in long-term erosion of a stream? Hazardous waste products that may contaminate water resources or pollute the air can be discharged either in the short term or over the life of the project. Waste discharges may damage an aquifer or the atmosphere. There are no boundaries for the distribution of atmospheric pollution, so global ecological consequences need to be identified and considered.

5.8.4 ADDITIONAL EVALUATION CONSIDERATIONS

At first, the evaluation of sustainability may appear to be easy: ensure that the natural resource in question is not depleted. This statement views sustainability as a single-criterion issue, which it is not. At a minimum, economic, social, and political issues must be considered. Other resource issues may also be factors in the evaluation. For example, if a second natural resource is drastically depleted in order to sustain the original natural resource, then the overall quality of a country's natural resources may degrade, which should be a factor in the evaluation. Additionally, risk and uncertainty must be addressed in the evaluation of sustainability. For example, an evaluation should consider the potential effect of a natural disaster on the sustainability of the natural resource. In addition, recognizing that special interest groups will be very vocal about sustainability adds to the complexity of the evaluation. The evaluation of sustainability is always a multicriterion activity.

Regardless of the criteria used to evaluate sustainability, the validity and reliability of each measurement must be considered:

- *Validity*: Does the evaluation metric selected accurately measure what it is intended to measure?
- *Reliability*: Would the evaluation metric provide the same assessment if it were applied either several times or by several different evaluators?

These measurements should be applied to both quantitative and qualitative evaluation criteria.

Many criteria are available to evaluate sustainability, and the importance of each will vary with the issue being evaluated. Potential evaluation criteria include:

- The natural resource is not excessively depleted during the evaluation period.
- Auxiliary natural resources are not sacrificed or adversely affected.
- The economic impact of sustaining the resource does not cause financial hardship to the community.
- The social effect of sustainable use of the resource is acceptable.
- Common resources are not overexploited, e.g., the "tragedy of the commons" effect.
- Risk and uncertainty analyses are conducted with the results considered in the evaluation.
- Political controls on the use of the resource are neither excessive nor lacking.
- Future trends in the use of the resource and in related resources are evaluated.
- Future projections suggest that the related goals will be met at that time.
- Critical milestones for future conditions are considered.
- Stakeholder concerns are adequately addressed in an unbiased manner.
- Consensus was reached on the establishment of priorities of each evaluation criterion.

The importance of each criterion will vary with the situation, and the weight applied to each criterion should be agreed upon prior to completing the evaluation.

5.9   KNOWLEDGE, SKILLS, AND ATTITUDES

Sustainability is unlike subjects such as solid or fluid mechanics, where a few basic principles can provide a strong foundation for solving relevant problems. Instead, the ability to apply sustainability requires a strong understanding of a wide array of topics, from economics to ecology, from psychology to public policy, and from soil science to systems theory. For this reason, single courses devoted to sustainability require a broad knowledge that may be better learned through a series of introductory courses in various disciplines. Then, a capstone course on sustainability that focuses on the application of the fundamental knowledge, skills, and attitudes can tie the individual components into a single focus.

To illustrate the breadth required to fully appreciate the meaning and application of sustainability, Tables 5.1 to 5.3 list a few knowledge, skills, and attitudes necessary. These lists are not intended to be complete. Instead, they should demonstrate the broad base of knowledge that an engineer needs to adequately address sustainability.

TABLE 5.1

Knowledge Related to Sustainability

Physical processes of natural systems

Public policy/political constraints and legislative issues

Life cycle assessments

Methods of waste minimization and renewable resource maximization

Effects of human activities on ecosystems

Benefit, cost, and interest calculations for engineering projects

Risk management

TABLE 5.2

Skills Related to Sustainability

Oral and written communication skills

Understanding interactions in ecosystems

Environmental impact assessments

Social and community impact assessments

Assessment of long-term effects

Quantitative data analysis and statistical methods

Risk reduction and management

Balancing competing project requirements

TABLE 5.3

Attitudes Related to Sustainability

Human values need to be considered in decision making

Decisions should be unbiased and objective

Ethical responsibilities are important

Balancing competing project requirements may be necessary

Public policies characterize public values and needs

5.10 DISCUSSION QUESTIONS

1. Identify ways that college students can promote sustainability.
2. Discuss public values relevant to sustainable development.
3. Identify reasons why research about sustainability is important to the design engineer.
4. Why is sustainability important to current generations?
5. Discuss the meaning of sustainability relevant to public welfare.
6. What physical processes are relevant to ecological sustainability? Define these processes and briefly discuss how they relate to civil engineering design.
7. Discuss the conflict between currently using natural resources to meet our needs and their preservation for the use of future generations.
8. One side often argues the following: The next generation will have sustainable energy sources (e.g., solar, wind, and wave power), so they will not need nonrenewable petroleum supplies. Therefore, we do not need to drastically reduce our use. Provide a counterargument to this viewpoint.
9. If we could go back in time to the development of asphalt roadways, what issues of sustainability would have been relevant?
10. Identify and discuss ways that the design engineer can promote sustainable development objectives.
11. Provide examples to illustrate how a construction contractor could promote sustainability on a construction site.

12. Identify and discuss ways that the owner of an engineering company can promote sustainability.
13. The automobile initially replaced public transportation facilities. Analyze this change from the perspective of sustainability.
14. Identify and discuss a policy that promotes sustainability.
15. Best management practices are one current solution to environmental sustainability. What new research about BMPs is needed to promote sustainability?
16. Identify reasons why research about sustainability is important to the design process.
17. Define LEED certification.
18. Compare sustainable home design with current home building practices.

5.11 GROUP ACTIVITIES

1. For one of the issues listed below, develop a plan that accounts for sustainability in solving the problem:
   - The transformation of a gasoline-powered transportation nation to one based on electrical-powered vehicles.
   - Landfills use valuable land, cause aesthetical problems, cause community problems, are generally viewed negatively, and are costly.
   - Dams provide many benefits, including power generation, recreation, and flood control. However, they damage fish spawning grounds, greatly alter sediment movement, and pose a failure risk.
   - Wetlands are drained to use the land for development, which can cause negative ecological changes.
   - The design of a footbridge for a location on campus.
   - The elimination of gasoline-powered vehicles forms the control park of a campus.
   - The use of green roof designs for campus buildings.
2. Identify the major energy uses in tall commercial buildings, such as those common to city centers. For each use, identify ways that the building could be transformed into a zero-energy building.
3. Small streams and the associated aquatic life are controlled by a number of physical processes. Stream restoration activities are intended to return a damaged stream to one that has a more natural balance of the physical, hydrologic, and ecological processes. Identify processes and their underlying scientific and engineering principles that are relevant to stream restoration.

REFERENCES

ASCE Code of Ethics, 1966, http://www.asce.org/inside/codeofethics.cfm.
Brundtland, R. *Our common future, Report of the world commission on environment and development*, World Commission on Environment and Development, 1987. Oxford, UK: Oxford University Press.

6 Contemporary Issues and Historical Perspectives

CHAPTER OBJECTIVES
- Present technology as a value input source
- Identify forces of knowledge—past, present, and future
- Show societal issues that have historically influenced engineering (and science) practice
- Discuss ways that civil engineers are involved in solving important societal problems
- Evaluate potential roles that civil engineers play in contemporary issues

---

6.1 INTRODUCTION

- How did World War I alter society's view of transportation engineering?
- Why were U.S. engineers unsuccessful in reducing boiler explosions in the early years of steam engines, i.e., the first half of the nineteenth century?
- Why are current design engineers indebted to Francis Bacon (1561–1626)?

The practice of engineering has historically been affected by the forces of society. Simultaneously, engineering and technology have influenced the direction taken by society. Consider the following:

- How did the development of machinery during the industrial revolution influence city planning and urban government?
- How did the fieldwork of John Wesley Powell influence the growth of environmental and associated public policies?
- How did Archimedes and his advancements in technology change the philosophy of society on the value of applied science?

Society affects the practice of engineering, and engineering is a force that influences the direction of society. This was true in the past, and it is true in the present, and the synergistic effects will likely be greater in the future. The interdependence of society and civil engineering places a demand on civil engineers to understand and appreciate history as an integral part of engineering. An understanding of historical forces will help engineers place contemporary issues in perspective.

6.2   TECHNOLOGY AS A VALUE SOURCE

Bremer (1971) identified five traditional value input sources that were found in a person's environment of recent generations: the agricultural society, family, town and community, religion, and education. Bremer hypothesized that while these were once the source of our values, a new source of values is replacing those of the past. The traditional mix of value input sources was time and space dependent. However, the growth of a technologically oriented culture, with its increased mobility and affluence, has altered the impact of each of Bremer's value input sources. Although Bremer contended that it was the influence of business that altered the source values, a reasonably good argument could be made that technology has been partially responsible. A better correlation exists between changes in technology and values than between changes in business practices and values, but this is largely a subjective observation. Technology currently has a significant influence on the formation of value systems. If engineers and scientists are largely responsible for much of technology, then it is not unreasonable to conclude that they may have a significant influence on the evolutionary path of human value systems. But it is not just technology that influences behavioral development; other forces play a significant role, too.

6.2.1   Is Technology Responsible?

Even though advances in technology have helped improve our way of life, technology has been blamed for many social problems. Because of the primary role that engineers and scientists have played in the growth of technology, they have been criticized for their role and blamed for some of the ills that have resulted from technological advances. But are technologists the force responsible for the evil side of technology? A solid case can be made that the responsibility must be shared by many elements of society, including scientists and engineers, courts and judges, politicians and governmental institutions, economists and managers, and last but not least, the public. For those who are dissatisfied with the results of technological growth, it is foolish to blame some impersonal, uncontrollable force called technology rather than the values and social structures that are responsible for the past allocation of both human and material resources (Florman, 1976).

6.2.2   The Forces of Knowledge

The examination of historical events should not be viewed as a series of dates and discoveries. The forces that encouraged or constrained the advancement of knowledge should be a focus of study. Knowing that Fleming in the early twentieth century "discovered" penicillin or that Coulomb suggested the use of compressed air in caissons in 1779 may be interesting, but it is important to know the societal forces that at points in time have severely constrained the growth of knowledge (Ziman, 1976).

Societal forces have also played a positive role in the development of knowledge. For example, in the sixteenth and seventeenth centuries, wealthy patrons made scientific advancements possible by providing talented scientists with the support needed to devote their intellectual energy to continue their research. For example, John Ray's (1628–1705) patron made it possible for him to study plant classification. The formation of the Royal Society of London in 1662 provided the institutional support for overcoming the Aristotelian forces that limited the growth of science in the seventeenth century. Knowing the context in which science and engineering advanced over the centuries is important because a new array of forces will control the expansion of engineering knowledge in the future.

In the past, the knowledge dimension of engineering has been constrained or encouraged by numerous forces, such as economics, theology, politics, and philosophy. The future growth of engineering knowledge may be constrained or supported by many of these same forces but also by new forces. Certainly, environmental constraints may be significant, which would be a change because environmental damage was not a constraint before the nineteenth century and possibly even much of the twentieth century. Politics has been a force throughout history and will continue to be in the future, but perhaps in different ways. Because of issues related to land and natural resources, governmental organizations, such as the U.S. Geological Survey, were created. The U.S. government directly affected research through the support of schools of forestry in the late 1800s. This research greatly influenced the use of the vast forests of the western United States. Just as politics can influence the growth of engineering and science, the growth of technology has had significant influence on the political arena. The industrial revolution was a principal force in the rise of many cities and created a need for governments that could address the many social ills that arose along with the technology. Events of the industrial revolution influenced the growth of city planning. We can ask ourselves, "What technologies of the future will cause a need for the development of new governmental forms?"

Religion was a dominant force on the growth of science and technology in the sixteenth and seventeenth centuries. We know that Copernicus's (1473–1543) hypothesis that the planets orbited the sun, rather than the Earth, was considered a radical theory by theologians. Earth was the domain of man, and the theologians believed that human beings should be the central focus. Galileo spent the last years of his life under house arrest because his views conflicted with accepted religious beliefs. Religion is still a dominant force in the acceptance of some new technologies and ideas (e.g., stem cell research, cloning, global warming, population control). Religion will likely continue to be a factor in the growth of technology and its engineering applications.

6.2.3 ANTITECHNOLOGY FORCES

Some groups want to tighten the reins on the growth of technology. They justify their views by pointing to the failures of past technologies. The use of nuclear power has been constrained partly because of the bombs and partly because of events such as Chernobyl and Three-Mile Island. The *Exxon Valdez* oil spill is cited as a concern for additional oil exploration and production on Alaska's North Slope. Antitechnology attitudes have always been a part of every society. The Luddites resisted change

CASE STUDY

The Luddites were one of a number of groups that opposed progress through technology. These groups were concerned with the loss of jobs and the replacement of skilled workers by unskilled labor. The mechanized looms introduced into textile manufacturing increased production rates and could be run by cheap labor. Thus, antitechnology groups conducted clandestine raids to destroy the looms in wool and cotton mills. The marauding began about 1811 in Nottingham, England, and it had such an effect on the economy that laws were enacted to make machine breaking a crime.

Obviously, the perception of technology has changed over the last two centuries, as technological change is now viewed more positively. The demand for more sophisticated communication devices is just one example. Society now expects engineers to solve our energy problems as well as improve our infrastructure. The social view of technology is an encouraging economic environment for engineering, much more so than during the time of the Luddites.

by destroying machines because they feared the type of society that was developing as industrialization progressed. However, their ability to stem the tide of technology failed, although laws and public policies developed to place constraints on the trend of the increasing use of machines. The enactment of child labor laws was one example.

6.3   HISTORY AND SOCIETY: DETERMINANTS OF A PROFESSION'S GROWTH AND DIRECTION

Heredity is believed to be an important determinant of an individual's value system. Similarly, history serves as a corresponding determinant in the formation of a profession's value system. Just as heredity is believed to limit an individual's value capacity, history in the form of social choices has had a significant limiting effect on the value capacity of a profession.

Just as environmental factors interact with the genetic determinants to mold an individual's value system, society interacts with historical determinants to mold the value system of a profession. Corresponding to an individual's value input sources, societal factors such as the following have been influential in shaping the value system of engineering and science professions:

- The expanding role of government in establishing public policies that control technological growth
- The need for organizational changes that will keep engineers and scientists from being separated from the moral consequences of their work
- The expanding gap between technological growth and the public's understanding of it

Historical events and the social choices made by past generations have influenced the state of professional values and engineering practice today. Three examples that illustrate the effects of events and social choices of the past on engineering practice today are

- Francis Bacon, a "scientist" around the time of the birth of modern science, developed a philosophy of scientific investigation that redirected the path of science and engineering.
- The loss of life due to explosions of steam engines in the early nineteenth century demonstrated the importance of an appreciation for public policy development.
- Resource policies and the social forces that shaped them exemplify the role of engineers and scientists of past generations in balancing economic growth and human welfare.

These three examples are discussed in the following sections.

6.3.1 The Birth of Modern Science

Francis Bacon (1561–1626) and René Descartes (1596–1650) were two major figures who stood at the turning point between medieval and modern science. Bacon has received more recognition for his role because he preceded Descartes and because the medieval system of thought was less entrenched in England than in France (Bernal, 1971). The scientific climate that existed prior to the sixteenth century had a significant limiting effect on the advancement of science and on the relationship between science and human values. Thus, it is of interest to examine the scientific climate that existed prior to Bacon, and note how his attitudes allowed him to be successful in changing the direction of scientific inquiry.

6.3.1.1 The Pre-Bacon Scientific Climate

The birth of Christianity represented a drastic change in value input sources. Paganism was replaced by a value system that put man above nature (White, 1967). However, an element of mythology remained. Instead of worshipping the sun, fire, and pagan gods, belief in a single god became the source of values. In addition, saints identified by religious leaders influenced the behavior of individuals. Science was compelled to practice within these constraints.

The scientific period before Bacon was greatly influenced by the humanists and Renaissance "Platonists," who interpreted natural science so that it maintained a proper harmony with the revealed dogmas of Christianity; that is, they served as defenders of past tradition rather than as initiators of empirical scientific research. During the fifteenth century, the literary and artistic efforts that were characteristic of the Renaissance period absorbed a considerable portion of the time of the best thinkers; the scientific investigation of nature received scant attention. Even into the sixteenth century, the humanists attempted to suppress development of the scientific method by questioning the value of studies of natural science. Erasmus (1466?–1536), like Petrarch (1304–1374) and Boccaccio (1313–1375) before him, implied that it was impossible to understand the natural world through empirical science.

The only writings in the fifteenth century that emphasized the importance of experience and experimental observation were the early works of Leonardo da Vinci (1452–1519); however, although he recognized the value of experimentation, he did not provide any systematic discussion of the scientific method. Leonardo's work is evidence that he had a good understanding of the scientific method. Not until the time of Galileo, however, did the scientific method become systematized.

The scientific climate of this period is aptly illustrated by developments in astronomy. Ptolemy (second century A.D.) argued that the simplest plausible theory was the most acceptable; the Ptolemaic system was accepted by many until the fifteenth century. Up to the sixteenth century, astronomers did not believe that empirical evidence was necessary to formulate or prove a theory, and they believed that failure to fit observed measurements was not a sufficient reason to reject a theory. Copernicus (1473–1543) was one of the first to require agreement between a theory and observed phenomena; however, he had to endure constraints on his freedom to pursue knowledge because of his unconventional approach to scientific investigation. Social forces placed control on the growth of knowledge.

The constraints on scientific advancement were compounded by theological factors. Copernicus was censured by the Holy Congregation of the Index because he presented his theory as a physical reality, rather than just a mathematical hypothesis. The movement of celestial bodies was considered to be a theological phenomenon, and theologians did not believe that astronomers were justified in examining theological phenomena. As a further constraint, failure to fit the astronomical teachings of the sacred scripture was considered sufficient grounds for rejecting a theory. Copernicus's curiosity and confidence in his experimental observations enabled him to overcome the theological constraints.

Understanding the scientific climate that existed before Bacon and Descartes and its effect on the development of the scientific method is important. The scientific climate also influenced both the evolution of professional values and the relationship between technological growth and human values. The theological constraints directed scientific investigation away from the physical world, which limited the ability of science to serve society. It was an atmosphere that limited the influence of religion as a value input source to professional development. Science could not develop as a means of serving human welfare; thus, it served only to tantalize the intellect of a few. Technological growth was constrained because religious beliefs discouraged people from engaging in scientific research. It was necessary to overcome this climate before scientific inquiry could proceed properly and before professional values could evolve.

6.3.1.2 Francis Bacon and the Scientific Method

Francis Bacon and others of his era sensed a change in the existing attitudes toward science and seized the opportunity to redirect the methodology by which science advanced. Because of his interest in both scientific inquiry and philosophy, he was quite aware of the link between science and human values. Bacon set much of human knowledge on a new path by (1) freeing science from the practices and privileges of the learned, and (2) separating truth that is humanly discoverable from the dogmas of theology (Anderson, 1971).

Bacon's views on knowledge were in direct opposition to the prevailing Aristotelian philosophy. The key to Bacon's doctrine was his emphasis on inductive investigation, which he detailed in *Novum organum*, dated 1620. Inductive investigations begin with particulars and proceeded to definition; Bacon's philosophy thus emphasized the value of experimental investigation and the identification of truth from the data obtained from observation.

Bacon's inductive approach to scientific investigation was an important factor in technological progress for several centuries, and his contribution to human welfare was just as important. Bacon preached the doctrine that the true and lawful end of the sciences is that human life is enriched by new discoveries. That is, the advancement of knowledge should be utility oriented, with the aim of relieving the burdens of life. This new philosophy had a profound impact on society and was used as a model for guidance by much of the scientific community. This in itself represents a significant change in the relationship between science and religion, which contributed to the change in attitude about science and human welfare. The philosophy and methods proposed by Bacon emphasized the importance of scientific investigation for the benefit of society.

6.3.2 The Steam Engine and Professional Values

The growth of the steam engine provides a historical example of the concern of engineering groups with the social impact of their technology and their relationship with public policies. The steam engine played a significant role in the expansion of the United States during the early nineteenth century. However, the high-pressure, noncondensive steam engine was an example of uncontrolled technology, which was indicated by the significant number of fatalities that resulted from boiler explosions.

Burke (1966) detailed the impact of this problem on federal power. In the early 1800s, the city council of Philadelphia was the first legislative body in the United States to recognize the problem and initiate an investigation. A group of engineers recognized the importance of experimental observation and recommended initial testing and regular proof tests to ensure the safety of all boilers. In June 1830, the Franklin Institute of Philadelphia recognized that both the engineering problems and the public policy constraints had not been resolved, so they empowered a committee to conduct experimental investigations to develop the knowledge necessary to formulate regulatory legislation. In addition to following the scientific method that evolved during Bacon's era, the committee also recognized the need to consider public safety; that is, the group believed that public safety should not be endangered by private negligence and a lack for knowledge. This concern reflected Bacon's philosophical doctrine that scientific advancement should reflect upon human values.

Unfortunately, it was not until the mid-nineteenth century that engineers were successful in getting legislators to enact laws that reflected their concern for public safety. It took the engineers and scientists almost fifty years to effect significant changes in public policy. The professional committees emphasized the values of life, security, and public health and safety, but they failed to appreciate the competing value, freedom, which was embedded in the economic system of the early nineteenth century. Just as the engineers and scientists of the current generation have had difficulty influencing public policy, the same difficulties limited the engineers' effectiveness 150 years ago.

Attitudes influence success, which is as true of professions as it is individuals. In the case of the steam engine, the engineers were curious about the factors responsible for steam boiler explosions. Their knowledge of experimentation enabled them to identify causes of the explosions. Most importantly, the profession was persistent in seeking changes to public policies. The engineers involved were confident that their knowledge could solve a significant societal problem, but their success was constrained more by the emphasis that society placed on free enterprise, which was a factor that the engineers initially undervalued.

6.3.3 NATURAL RESOURCE POLICY AND THE PUBLIC INTEREST

The development of a natural resource policy in the United States is another example of the engineering profession applying both Bacon's scientific method and his concern with the public interest. John Wesley Powell is credited with being a pioneer in the development of federal science (Wengert, 1955). At a meeting of the Geological Society of America during the last decade of the nineteenth century, Powell identified two stages in development: a preliminary experimental or preparatory stage and the final or effecting stage. During the first stage, methods are devised, experiments are conducted, scientific apparatus is invented and subjected to trial, and a plan for the experimental design is formulated. During the second stage, the methods and apparatus are employed practically and the plans are carried out. Powell stated: "It is the highest function of systemized knowledge to promote human welfare, ... the first stage represents the seed-time, the second the harvest-time of science" (Rabbitt, 1969, 24). It is especially noteworthy that this philosophy directly parallels Bacon's two-phased philosophy, with emphasis on human welfare in the second phase. John Wesley Powell applied this philosophy throughout his career as an engineer and statesman. It represented the concern for human welfare that was characteristic of the engineers of his time. Powell's major emphasis was on the use of his observations in the development of public policy that would ensure progress for the human race.

Unfortunately, Powell was dismissed from his post as secretary of the interior in the late nineteenth century because of his insensitivity to political constraints and in spite of his logical program formulation of national resource policy (Rabbitt, 1969). This parallels the frustration of the engineers investigating the steam boiler explosions in the earlier part of the nineteenth century. Powell failed to appreciate the expansionary forces that existed in the late nineteenth century, and he failed to understand the use of political power and the potential of professional societies to affect public opinion and public policy. It wasn't until many of his followers, such as Gifford Pinchot and Teddy Roosevelt, became influential in the political process that many of Powell's recommendations on land policy and the role of government in science were institutionalized. Powell had a good value system, as evident from his concern for the sustainability of natural resources, but his lack of appreciation for political constraints limited his success.

6.4 ARCHIMEDES AND BERNOULLI: LEADERS IN THE ADVANCEMENT OF KNOWLEDGE

Those who have taken a course in the fundamentals of fluid mechanics are quite familiar with the names Archimedes and Bernoulli. Archimedes (287–212 B.C.) is the central figure in the story of the Greek scientist who jumps out of the Roman bath and runs down the street while yelling, "Eureka! Eureka!" which translates to "I've got it!" As the story goes, a flash of insight came to Archimedes while he was in the public bath, and it led him to the discovery of a nondestructive method for determining whether or not King Hieron's crown maker had pilfered some of the king's gold in making the crown.

Daniel Bernoulli (1700–1782), who followed two millennia later in the footsteps of Archimedes, is of interest to us because of his achievements in hydrostatics. Specifically, Bernoulli's equation, which defines the relationship between the pressure, velocity, and elevation heads, is an important tool in the solution of pipe flow problems. However, an interesting point is that Bernoulli did not ever propose or use the equation that bears his name. Bernoulli was only concerned with the relationship between the pressure and velocity heads. Only later was the elevation head inserted by others into his equation.

Archimedes and Bernoulli lived and practiced in very different periods of history. Archimedes lived a century after the golden age of Greece, while Bernoulli lived a century after the closing days of the Renaissance. In spite of the two millennia difference between the times of Archimedes and Bernoulli, many of the same forces that affected their contributions to the water sciences are evident. Both Archimedes and Bernoulli had to overcome cultural constraints and the lack of scientific knowledge to provide monumental advances in the water sciences that endure to this day.

6.4.1 MATHEMATICS AND ADVANCEMENTS IN KNOWLEDGE

Archimedes is known because of his seminal works in the field of hydrostatics, most notably in buoyancy. However, mathematics was his true interest. Some have argued that Archimedes was actually the father of calculus, the person who paved the way for Leibniz (1646–1716) and Newton (1642–1727). Archimedes used the concepts that underlie integration to compute the area of parabolic segments and the volumes of spheres and other solids. He was also a military strategist and was involved in the design of instruments of war. He used his exceptional understanding of mathematics to create catapults and other machines that held off the Roman siege of Sicily for almost three years.

Archimedes was not a mathematician in the sense in which we currently categorize mathematicians, but even then, mathematics was a core discipline of knowledge. In his time, mathematicians were considered to be philosophers who used mathematics and the methods of science to create order in their thinking. A few found mathematics to be a useful tool to achieve their goal of solving problems, whether it was to assess the amount of gold in the king's crown or to develop a hydraulic screw for lifting water to irrigate fields.

Coming from a family of mathematicians provided Bernoulli with the atmosphere in which he could develop a strong mathematical foundation. He sought to follow in the philosophy of both Archimedes and Francis Bacon and provide advancements in applied science. In his study of gases, Bernoulli introduced concepts of uncertainty and probability, which were just beginning to be developed by Pascal, Fermat, and Bernoulli's uncle. This acceptance of mathematics as part of scientific advancement enabled Bernoulli to provide a mathematical framework for significant advancements in the fluid sciences.

6.4.2 ADVANCES IN THE METHODS OF SCIENCE

Another commonality between Archimedes and Bernoulli was that they followed monumental advances in scientific thought proposed by preeminent names in the advancement of science, namely, Aristotle and Isaac Newton, respectively. Aristotle (384–322 B.C.) was one of Plato's (427–347 B.C.) students but rejected Plato's teachings that experimentation was a crude method for scientific advancement (Durant, 1939). While Aristotle was not an experimentalist, he did show that keen observation and logic could advance scientific knowledge. This philosophy served as the model for Archimedes' advancement of the method of science, most notably the importance of experimentation (see Chapter 4).

In addition to his advances in mathematics, hydrostatics, and statics, Archimedes wrote on the method of science. The greats of "modern" science at that time included Newton, Descartes, Bacon, and Galileo. They had the works of Archimedes from which to learn. Kearney (1964) indicated that ancient scholars correctly formulated only three physical laws: the optical law of reflection, the principle of the lever, and the principle of buoyancy. In spite of the many great scholars of that period, one person, Archimedes, was solely responsible for two of these physical laws. His development of the lever and the idea of the center of gravity shows that Archimedes is the father of the science of statics. Many people consider Archimedes' greatest achievement to be his writings that paved the way for the advancements of science beginning in the sixteenth century. Galileo credited Archimedes with much of his philosophy of science (Kearney, 1971). Archimedes' redirection of the method of science toward experimentation enabled him to make his advancements in and direct the research of the great scientists who followed him, such as Galileo and Francis Bacon.

Beginning with Francis Bacon in the early part of the seventeenth century, the scientific community moved away from Aristotelian thought and adopted a method of science that more closely aligned with Archimedian investigation, specifically analysis and synthesis. The works of Archimedes had been lost to most scholars until the sixteenth century, which allowed the time for Aristotelian science to become entrenched. With its deemphasis on experimentation, Aristotelian science was less likely to lead to general principles with broad application. Archimedes thought that experiences could be analyzed to discover basic principles. Deductive reasoning, or synthesis, would uncover the solutions to practical problems. This was the foundation for the seventeenth-century advancements in the method of science.

Following the philosophical ideas of Francis Bacon and René Descartes (1596–1650), Isaac Newton (1642–1727) proposed a method of scientific investigation based on the belief that advances could be made by observation followed by the formation of hypothesis and the collection of data. Robert Boyle (1626–1691) observed changes in the pressure of gases with varying temperature but had not mathematically defined the relationship. Bernoulli was interested in the behavior of gases under pressure. Because of his abilities in mathematics and experimentation, Bernoulli provided the relationship that Boyle had sought. Bernoulli followed Newton's method of scientific investigation to extend this knowledge to the water sciences. From this knowledge he developed the initial form of Bernoulli's equation that would indicate that the sum of the pressure and velocity heads is constant.

6.4.3 PROFESSIONAL COLLABORATION

Communication among professionals can be a major factor in the advancement of knowledge. Both Archimedes and Bernoulli benefited from professional collaboration. While the scientific center in Alexandria, Egypt, had lost some of its prestige over the course of the third century B.C., Archimedes, who lived in Sicily, interacted with its mathematicians and scientists. While communication was limited by the existing modes of travel and the venues for writing, ideas and philosophies of scientific investigation were exchanged regularly among the philosophers and mathematicians of the day. This collaboration contributed to Archimedes' success. This was an example of globalization of science, third century B.C. style. These professional interactions are the forerunners of the teamwork that dominates professional life in the twenty-first century.

The positive climate for advancement in the methods of science in the seventeenth century provided an atmosphere that encouraged scientific interaction. This contributed to the development of professional societies, such as the *Accademia dei Lincei* in Italy (1603), the Royal Society of London (1662), and the *Academie of Sciences* in Paris (1666). These scientific organizations were important both in overcoming many of the cultural constraints and in the advancement of the new methods of science. Most noticeable was the replacement of Aristotelian dogma with the more systematic methods of Bacon and Newton. Such scientific organizations provided scientists with both the collegiality and the competition that spurred advancement. The formation of these scientific academies marks the seventeenth century as the beginning of science as an organized social activity (Ziman, 1976). The momentum initiated in the seventeenth century carried over to the eighteenth century, all of which helped to smooth the way for Bernoulli's developments.

6.4.4 CULTURAL INFLUENCES

The influence of Greece declined during Archimedes' lifetime. Along with this went a decline in social stability. War became a reality with which the scientists had to contend. Rome was becoming a more powerful force in Sicily. While King Hieron of Sicily initially sided with the Romans, eventually disagreements arose and Rome attacked Sicily. Archimedes spent considerable effort in developing machines of war. This acted as a constraint on his mathematical and scientific investigations, except for those that were related to the military defense of Sicily. In the second century B.C., society was influencing the conduct of science and engineering.

During Archimedes' lifetime, slaves were the principal source of labor, so scientists had little motivation to develop labor-saving technologies. This cultural attitude influenced the direction of scientific study. Efforts toward the development of machines were not highly respected because of the supply of slaves. This discouraged Archimedes' scientific inquiry into machines, but in spite of this, he made significant advances, especially with respect to the simple lever. This illustrates the point that social conditions can affect professional practice just as professional practice can affect social change.

Just as social conditions affected Archimedes' practice, Bernoulli had to contend with social constraints. Bernoulli's advancements of knowledge may seem trivial to us, but it is important to consider the society and everyday life that existed in his time. He practiced a century after the Renaissance, when it was more common to study the historical authors than spend time advancing knowledge. The works of Aristotle were readily available; scientists studied his works even into the seventeenth century. Thus, the beliefs of Aristotle directed the research of many. Any attempt to advance knowledge was viewed as being disrespectful to the authorities of the previous generations. Furthermore, Bernoulli started his investigations only about a century after Bacon had introduced a methodological approach to scientific investigation, which had not been fully accepted by scientists and mathematicians. These cultural attitudes acted as constraints on Bernoulli's practice of investigation.

Looking beyond the climate within the scientific community, the social, religious, economic, and intellectual behavior patterns of the time acted as constraints in some ways and as stimuli in other ways. In the sixteenth and seventeenth centuries, the practice of witchcraft was quite common. Astrology was a force in everyday life. Noted scientists like Tycho Brahe (1546–1601) and Johannes Kepler (1571–1630) practiced astrology. Even as late as 1759, which was the year that Edward Halley (1656–1742) had predicted would be the next return of the comet that bears his name, many scientists and philosophers were superstitious, such that they believed comets were bad omens. Alchemy was influential as late as the seventeenth century. Alchemists argued that they could turn base metals into gold and develop elixirs for longevity. Since riches and longevity were desired by kings and princes, those who dabbled in alchemy were protected. These beliefs and practices limited the advancements of scientific thought and influenced the practices of science even into the start of the eighteenth century.

6.4.5 A CONCLUDING THOUGHT

Several hundred years from now, when the lives of famous engineers of the twenty-first century are being written about, what factors will be cited as those that encouraged advancement of engineering and what constraints will have limited advancement? Quite possibly, the computer or one of its offspring will be identified as a major factor for advancement. This effect might be similar to the effect that advancements in mathematics had on both Archimedes and Bernoulli. Advances in instrumentation, such as the ability to detect miniscule concentrations of pollutants, may be a factor, much as advances in instrumentation played a role in the times of Daniel Bernoulli. The Internet and globalization may encourage collaboration of engineers separated by thousands of miles, much as Archimedes' associations with the Egyptians and the Greeks. The entrepreneurial spirit that encourages research, much like the attitude toward Francis Bacon's method of science, may also be cited as a positive factor in the lives of the twenty-first-century engineers who will make their mark. This would be similar to the enlightened attitude about scientific advancement in the time of Bernoulli.

Just as Archimedes had to overcome philosophical constraints prevalent in the third century B.C., and Bernoulli practiced under the theological constraints of the eighteenth century, engineers of the twenty-first century may be limited as they attempt to advance knowledge. The lack of resources for research can be a very constraining factor. Religious attitudes that oppose scientific advancements, such as cloning, genetic research, and stem cells, can limit advances in knowledge. This parallels the theological beliefs that acted as constraints in the time of Bernoulli. Thus, parallels can be drawn between the practices of and constraints on the advancement of knowledge in the different millennia.

6.5 ENGINEERING INVOLVEMENT IN CONTEMPORARY ISSUES

The discussion of the historical forces illustrated several important points:

- Attitudes are important to success.
- Social forces affect the advancement of knowledge, and those active in advancing knowledge affect society.
- Professional values can be influenced by society.
- Engineers who lack an appreciation for public policies and attitudes may be less effective in affecting change (e.g., engineers involved in resource policy development).
- Engineers who do not appreciate economic forces (e.g., the engineers involved with the steam boiler explosions) have difficulty in having their knowledge respected.

Many of these factors are evident in contemporary issues. A few examples are briefly discussed in the following sections.

6.5.1 CONTEMPORARY ISSUE: THE IMPACTS OF TERRORISM

Citing instances of terrorism is much easier than providing a definition. *Terrorism* may loosely be defined as the use of violence to achieve an end. Terrorists use violence against people or infrastructure in order to intimidate a community into accepting the beliefs of the terrorists or to retaliate for some perceived wrong. Terrorism affects the civil engineering profession in many ways, the most physically obvious of which is when the act of terrorism is directed at infrastructure. Domestic terrorism, such as the Oklahoma City bombing, and international terrorism, such as the World Trade Center collapse, cause changes in the way that civil engineering is practiced. While terrorism often has its roots in politics and political causes, it effects are far reaching, including into the civil engineering profession. Even acts of terror committed in other countries can influence civil engineering practice in the United States.

The effects of terrorism can influence civil engineering practice, education, and research. As worldwide terrorism has been on a rapid rise, the demand for

---

CASE STUDY

Possibly the most recognized terrorist act on a ship was that on the USS *Cole*. The blast severely damaged the ship, and sailors onboard were killed. This act showed that ships were vulnerable to terrorist acts and engineers needed to identify countermeasures to protect ships. That has led to the design of floating barriers that are used to surround a ship. The technical basis of the design of these "fences" is quite simple, as Archimedes' principle applies: the weight of the barrier must equal the weight of the displaced fluid. The weight of the flotation barriers, the steel fence, and all connections is balanced by the weight of the fluid displaced by the submerged portions of the fence. The use of such facilities reflects the design engineer's expertise in problem solving.

---

engineering solutions has not necessarily allowed for the major changes in civil engineering education. Instead, individuals have retrained themselves through self-learning to address technical matters unique to the effects of terrorism. The need for civil engineering research related to the effects and control of bomb explosions has increased and refocused a number of research efforts. The need for new civil engineering research on the effects of blasts on building fires and loss of lifelines has also increased. The need for education and research will grow with the frequency and magnitude of terrorist acts.

The issue of terrorism clearly affects engineering practice, as the design engineer needs to consider new effects. For example, while fire had been an important building design consideration, the fire load created by the jet fuel in the World Trade Center case was beyond the normal design condition. Future building designs will need to consider this potential factor.

6.5.2 Contemporary Issue: Conflicts over Water Supply

Water is an important commodity even in humid areas, but even more so in arid and semiarid climates. If one state controls access to water rights, a second state may not believe that it has a sufficient supply. Political and armed conflicts have developed over water rights. The countries of the Mideast frequently hold meetings to discuss water rights and the sharing of the tight water supply in arid regions.

How can civil engineers reduce the need for conflicts over water? Better planning is the first step. Times of drought will occur, and planned water development projects can reduce the negative consequences of a drought. Second, public policies need to be developed to fairly allocate the resource between political jurisdictions or interest groups. Civil engineers who have knowledge of water supply can assist in developing these policies. Engineers who have knowledge beyond the technical issues will be more effective if they understand the process of developing public policies. If the civil engineer does not appreciate the importance of public discourse and policy development, then the technical issue may be sacrificed because of political issues.

Third, engineers can provide the leadership to facilitate compromise over allocation between the conflicting interest groups. A civil engineer who has strong leadership ability is better able to effect a compromise that will be both technically optimal and politically acceptable to the conflicting parties. Fourth, civil engineers involved in research can develop better ways to use the available supply of water, e.g., more efficient irrigation methods or methods of reducing losses, such as leaks from pipes. Water supply is important to every person worldwide, and it is a contemporary issue in which civil engineers will play a vital role.

6.5.3 CONTEMPORARY ISSUE: RECYCLING

Recycling around the home has been a common practice for only about a decade. Before that time, little thought was given to the implications of discarding trash. Energy consumption, landfill space requirements, and the depletion of natural resources were not considered economic or social problems by much of the public. Society's views on these issues have significantly changed over the last decade. Many paper and plastic products are currently made with recycled materials. Recycling reduces landfill space requirements and minimizes the use of new raw materials and energy. Thus, recycling is also a sustainability issue. Efforts to increase recycling are made at all levels of government and even within some organizations. In addition to paper waste, used water, often referred to as gray water, is recycled for use where clean water is not necessary. The economics of recycling has in the past limited its growth, but recent improvements in economies of scale have made recycling more cost-effective.

Engineers are becoming more involved in recycling. For example, when a building is torn down to be replaced by a new structure, the materials that were used in the old building are often salvaged and recycled for use in the renovation of other existing structures. In addition to reducing the use of raw materials, this recycling often reduces the energy used because the recycled materials require less transportation than the movement of new building materials to the site being renovated.

The construction of roadways offers another opportunity for using recycled materials. Recycled glass products as well as old concrete and car tires can be incorporated as aggregate into new roadways, thus reducing both the use of new raw materials and the space required for dumping the discarded waste. This has implications to natural resource sustainability.

Engineers can influence the recycling effort by directing the use of recycled materials in their design and construction practices. For many decades, soil has been recycled. Cut material was used as fill material at other parts of a construction site. While this is not a new recycling practice, increasing its cost-effectiveness is an important role for engineers.

6.5.4 CONTEMPORARY ISSUE: ECOLOGICAL CHANGE

While many human-induced ecological changes are occurring, the development of algal blooms in lakes and coastal waters is of special concern. Weekly news magazines occasionally report on the growth of the dead zone in the coastal waters around the mouth of the Mississippi River. Flows that originate from agricultural areas of the upper Midwest are laden with soil eroded from agricultural fields. Commercial fertilizers that include a considerable amount of nitrogen and phosphorus are used on the fields, and much of the excess is transported with the eroded soil down the Mississippi River. Without the fertilizers, the supply of phosphorus would not be sufficient to support large algae populations. One implication of this change to the ecology of the receiving water is that the increased algal population creates a habitat that can no longer support fish and other aquatic life, such as crabs that would normally be harvested for commercial sale. The loss of such organisms can affect the local economy. Our inability to solve the problem has both economic and ecological effects.

Engineers can play a significant role in both the prevention of the algal problem and reducing the effect of the problem once it begins. Specifically, engineers can work on the problem at the source, i.e., the agricultural fields; at intermediate points, i.e., the small streams that receive the sediment-phosphorus-laden storm runoff; and at the site of the problem, i.e., the receiving water body where the sediment accumulates and provides the environment where the algae can thrive. Engineers can also develop methods that reduce the detachment and transport of the sediment from the fields, including best management practices that can trap the material before it leaves the fields. Engineering research can identify ways to clear the receiving waters of the algal blooms once they have started. Solving this problem will allow affected receiving waters to recover and begin to provide for sustainable harvesting of the fish and crabs. Thus, in addition to the positive ecological effects, preventing or solving the problem will increase the availability of jobs in water-based fishing industries.

6.6 DISCUSSION QUESTIONS

1. Discuss how the rise of machinery during the industrial revolution changed class distinctions.
2. Review a couple of articles from professional journals that report research related to the effects of bomb blasts on infrastructure. Based on the papers, discuss the types of problems that could not be addressed without new research.
3. Navigation using the new astronomy of the Renaissance had a practical economic consequence. GPS currently has economic value. Compare and contrast the two.
4. Ages for stone, bronze, and iron are part of history. Nanomaterials are becoming a significant factor in the present economy. Discuss the effects of stone, bronze, and iron on economic growth and the potential of nanomaterials in the twenty-first century. Identify similarities in the roles played by each of the materials in their ages.
5. The industrial revolution saw a number of changes: from wood to coal as a fuel, from wood to iron as a material, from water power to steam power, and from single to multiple action in spinning. What roles did engineering and science play in these transformations?

6. Most of the discoveries of the sixteenth, seventeenth, and eighteenth centuries were the work of individuals. Since the time of Edison, invention has been largely teamwork. What social forces contributed to this change?

7. Analyze the current clash between capitalist forces and the socialist forces of the working class. Then compare this clash to that of the industrial revolution.

8. Analyze the economic and social forces at the times the plow was developed, the tractor became the workhorse of the farm, and the expansion of the current dominance of agribusiness.

9. Analyze the economic impact of the engineering wonder the Panama Canal, and project from the analysis how massive engineering works can affect the current economy.

10. Analyze the social impact of the effort to land men on the moon in the 1960s. Discuss how this engineering marvel influenced the public's opinion of engineering.

11. Engineering advancements have had significant economic, social, and environmental impacts on society. Discuss whether or not this has provided the engineering profession with power and influence on economic decisions.

12. Obtain estimates of the tax dollars devoted to science and engineering research over the last three decades. Discuss whether or not this is viewed by the public as money well spent.

13. Engineers of the 1850s understood factors that influenced the strength of iron and steel, but lacked knowledge of the mechanical properties. What historical forces led engineers to investigate the mechanical properties?

14. Eli Whitney (1765–1825) was the American inventor of the cotton gin. Discuss the economic, political, and societal impact of the technology.

15. Discuss the potential impact of robots on the U.S. economy and the environment.

16. Identify the potential effects of federal energy policies on the engineering profession.

17. Federal and state governments provide funds that support engineering research. How does engineering research affect society? Identify and discuss an example of engineering research that has had an economic or environmental benefit to society.

6.7 GROUP ACTIVITIES

1. In the early 1800s, engineers designed and constructed a number of canals, which had a significant impact on society, including economic benefits. The canals were influential in westward expansion. Identify a canal that was especially important to a region and detail its role in society.

2. Technology creates value dilemmas. Identify an example of technology associated with civil engineering and discuss value dilemmas presented by the technology.

REFERENCES

Anderson, F. H. 1971. *The philosophy of Francis Bacon*. New York: Octagon Books.

Bernal, J. D. 1971. *Science in history, Vol. 2: The scientific and industrial revolution*. Cambridge, MA: MIT Press.

Bremer, O. A. 1971. "Is business the source of new social values?" *Harvard Business Review*, 15, 121–126, Nov–Dec.

Burke, J. G. 1966. "Bursting boilers and the federal power." *Technology and Culture* 7(1):1–23.

Durant, W. 1939. *The life of Greece*. New York: Simon and Schuster.

Florman, S. C. 1976. *The existential pleasures of engineering*. New York: St. Martin's Press.

Gengart, N. 1955. *Natural resources and the political struggle*. New York: Random House.

Kearney, H. 1964. *Origins of the scientific revolution*. London: Longman, Green & Co., Ltd.

Kearney, H. 1971. *Science and change: 1500–1700*. New York: The McGraw-Hill Companies.

Rabbitt, M. C. 1969. "John Wesley Powell: Pioneer statesman of federal science," in The Colorado River Region and John Wesley Powell. USGS Professional Paper No. 669. USGPO, Washington, D.C.

Weingart, N. 1955. *Natural resources and the political struggle*. New York: Random House.

White, L., Jr. 1967. "The historical roots of our ecological crisis." *Science* 155(3767):1203–1207, March 10.

Wolper, L. 1992. *The unnatural nature of science*. London: Faber and Faber.

Ziman, J. 1976. *The force of knowledge*. London: Cambridge University Press.

7 Risk and Uncertainty

CHAPTER OBJECTIVES
- Define and present applications of uncertainty in both knowledge and data
- Provide a brief interpretation of engineering risk
- Discuss the triad of risk, technology, and public policy

7.1 INTRODUCTION

Consider the following statement from a building code:

> If a builder builds a house for a man and does not make its construction firm and the house which he has built collapses and causes the death of the owner of the house, the builder should be put to death.

Would you like to practice engineering under such a code? Fortunately, this was part of the Code of Hammurabi, which dates to about 1750 B.C. Given the potential hazards at that time and the lack of knowledge about materials, natural disasters, and construction practices, the life of a contractor was likely risky.

Risk and uncertainty are a part of life, both in our personal lives and in our roles as engineers. People in coastal areas are at greater risk of enduring the effects of a hurricane or a tsunami than those in the interior parts of a country. However, the latter are frequently threatened by tornados. Earthquakes, mudslides, and temperature extremes put people's lives at risk. Beyond risks associated with natural disasters, individuals and communities are subject to risks associated with acts of terrorism, hazardous waste spills, and traffic accidents. Events with significant risk influence the direction of engineering, and engineers can influence the magnitude of the risk associated with the consequences of an event.

Probabilities of risky events such as natural disasters are uncertain and often misunderstood by the public. Consider the following:

- Which is more likely—death from a fall or death from being struck by lightning?
- Which is more likely—death during construction work or death during a boating accident?

Responses to these queries are sort of obvious. Death from a fall or during a boating accident is more likely than the alternatives, but the disparity in frequency is not

NATURAL DISASTERS

The Galveston Hurricane of 1900 is classified as the United States' deadliest natural disaster, which killed more than 8,000 people. Although residents had pushed for a seawall to be built prior to the hurricane to protect against such storms, officials reasoned that a storm of high magnitude would never hit the island. After the storm, residents rebuilt, raised the sea level of the island from 8 feet to 17 feet, and constructed a seawall. In 1900 the island was a booming, prosperous town; however, after the hurricane, economic traffic was shifted away from the island, and the island's growth was stunted. The Galveston Hurricane of 1900 is an important historical lesson to engineers. Specifically, proper precautions must be taken even though they may be expensive. The event also serves to show the importance of understanding risk and uncertainty. In 1900, they had very little understanding of risks associated with natural disasters. Even today, we still do not have methods for accurately assessing risks, and with changing climate conditions, even the estimated risks are uncertain.

so easy to guess. Death by a fall is more than 100 times more likely than death by being struck by lightning. The chance of being killed on a construction site is less than 7% of the likelihood during a boating accident. Probabilities of engineering failures are often more difficult to estimate, which makes it difficult to evaluate the likelihood of damages, which are often the responsibility of engineers.

7.2   VALUES RELEVANT TO RISK

Human values are central to the civil engineer's role in risk. All designs, especially those associated with natural occurrences, are subject to risk. Therefore, values such as the following are central to the risks and uncertainties that engineers must address:

Life: Activities, relationships, and interests collectively.
Safety: Being free of danger, risk, or injury.
Welfare: The general well-being of individuals and the community.
Competence: Having the capacity and ability and to be qualified.
Knowledge: Understanding gained through study or experience.
Care: Being conscientious and attentive to detail.

Public life (or health), safety, and welfare are often listed as primary values in codes of ethics. Each of them have an obvious connection to risk. Risky activities, events, or designs pose a threat to the public, and the more effort that engineers make to reduce project risk, the greater the benefit to the public. Engineers recognize that all designs are associated with some element of risk. Small inlets that control flood runoff into storm drain systems sometimes are inundated, with local flooding the result. The risk of floodwaters overtopping a levee is certainly an engineering concern, as the potential damage is much greater than that of a small culvert. However, both projects carry risks that design engineers must consider in sizing the facility.

A design engineer who does not recognize and fully address risk and uncertainty is of questionable competency. Competency implies having adequate knowledge of potential risks and the capacity to adjust the design to minimize risks to public health, safety, and welfare. Engineers must take proper care to ensure that the public is not subject to unnecessary risk.

7.3 DEFINITIONS

While dictionary-type definitions could be given, the definitions provided below are intended to be applied to the issues of risk and uncertainty in civil engineering.

Uncertainty: A condition that results from a lack of knowledge or doubt about observations.

Knowledge: (1) Specific information that is known and is considered to represent the state of the art as agreed upon by experts. (2) Knowing or familiarity with engineering design methods.

Data: Organized information that is often the result of observing and practical experience.

Measurement: The act of ascertaining a characteristic (e.g., force, length, time, mass, concentration) of a property or variable.

Note that knowledge has two parts, the information itself and the degree of familiarity with it. Also note that uncertainty applies to knowledge, data, and measurement. In many cases, the three sources of uncertainty are not independent.

Examples

The following examples illustrate uncertainties in

Knowledge:
- The effect of global warming on the strength of hurricanes
- The effect of rust layer cracking on steel corrosion
- The reaction of people to imminent natural disasters

Data:
- Variation of evaporation from a lake
- Variation of compression strength of concrete cylinders
- Variation of dissolved oxygen concentration in a stream

Measurement:
- Multiple measurements in length between two survey points
- Reading on a deal when the visual angle is skewed
- The inability to read a graph, such as a Moody diagram

7.4   UNCERTAINTY OF KNOWLEDGE

Knowledge can be separated into two parts, that of the system and that of the user, i.e., design engineer. The former refers to the physical processes being analyzed, the inputs and outputs of the system and the design process, each of the variables in the process, and environmental factors, which are defined as processes and variables that are part of the system but are not part of the design method or model. The latter, i.e., user knowledge, may for any individual be limited and less than the knowledge that comprises the accepted state of the art. The sensitivity of the design method to various assumptions, inputs, and processes is an important part of the knowledge of a design method. For example, when a model is used as a part of a design, the user is assuming that the model is an accurate representation of the physical processes. The uncertainties used in developing the model are often not directly considered by the design engineer.

Theory often underlies civil engineering design methods; however, the direct transfer of theory to practice is uncommon. Very often, the theory must be calibrated because direct application may be inaccurate. The fitting of the theory with measured data reflects the degree of uncertainty in the theory. For example, a design model may be based on the assumption of homogeneity of materials or soil, while inhomogeneity is the norm. The lack of homogeneity must be accounted for through fitting values of coefficients. Even simple design tools like the rational method for estimating peak discharge rates includes an empirical coefficient (C) to relate the discharge (q_p) to the rainfall intensity (i) and the drainage area (A): $q_p = CiA$ In this example, the C value transfers the volume rate of rainfall, iA, over the watershed to a volume rate of runoff, q_p, at the outlet. Thus, the value of C reflects immediate losses due to infiltration, interception, and surface storage, as well as physical characteristics such as watershed slope and soil properties.

In addition to the inapplicability of underlying theory, uncertainty is introduced because of the knowledge of the user. The lack of a full understanding of the theory and the problems associated with its application can introduce uncertainty. For example, the rainfall intensity in the rational method depends on a computed value of the watershed time of concentration, where an inaccurate estimate can result from either an inaccurately measured input such as the slopes of the principal flow path or the roughness coefficients. If an engineer fails to maintain competency, design estimates made by the person will reflect a measure of inaccuracy in the estimated discharge due to the person's uncertainty in knowledge.

7.5   UNCERTAINTY IN DATA AND MEASUREMENT

Measured data can be inaccurate in two ways. First, does the measurement actually reflect the intended characteristic? For example, does Manning's roughness coefficient really measure the effect of retardance to flow? Second, does the measured value reflect conditions at the site? For example, if Manning's roughness coefficient was estimated in one gutter, would it reflect the roughness of all gutters within an urban drainage system? As another example, if samples of the total suspended solids in a river are taken once a month, are they valid indicators of the amount throughout the month?

The uncertainty in a measured value is indicated by its sampling variation. If millions of measurements were made of the plastic limit of a particular soil, then the variation would be known fairly precisely. However, when only a few measurements are made, the variation of the sample values would be the best estimate of the sampling variation or uncertainty of the property. The larger the sample, the better the estimated value, but the uncertainty still exists.

7.6 ENGINEERING RISK: INTERPRETATION AND ESTIMATION

While the idea of risk is rarely misunderstood, some confusion exists about how engineering risk is measured. In some cases, risk is represented by the probability of occurrence or nonoccurrence of an event, while risk is often converted to an expected value of monetary benefits or loss.

It should be emphasized that a risk exists only when a hazard exists and when something of value, including human life, is exposed to that hazard. To quantify risk, both a description of the vulnerability and a damage function are needed (Lee and Collins, 1977). A hazard can be either a natural or man-made phenomenon, and a hazard model, which provides a complete probabilistic statement of the hazard, is used in assessing the risk. The damage function provides a description of the losses or consequences for selected damage scenarios. Risk is then the product of the probability of occurrence and the consequences.

Scientists, engineers, and laypersons agree that risks associated with technological growth should always be identified. Siekevitz (1970) believes that the scientist has a responsibility to identify the social risks and merits as well as the scientific merits of research. However, he correctly adds that it is often impossible to identify the consequences. For example, the ecological effects were not evident at the time pesticides were first marketed and were only fully understood after years of use. Many individuals now argue that a greater share of research funds should be spent on the social effects of research results. Some individuals counter that because technological assessment of the long-term risks of a new product is quite often not fruitful, funding for risk assessment represents an inefficient use of resources.

CASE STUDY

DDT is a good example of a product of science that had both positive and negative effects. DDT was developed as an insecticide, and resulted in a savings of more than 5 million lives in the first decade of its use. For example, it was very effective in reducing the incidence of malaria in Ceylon. However, DDT has serious environmental effects. For example, DDT interferred with the life cycle of birds. Because of the seriousness of the risks involved to both humans and fauna, the United States banned DDT in 1972. It was felt that the environmental and health risks outweighed the benefits.

CASE STUDY: FLOOD RISK

It is a common misconception that if the so-called 100-year flood occurred last year, it will not occur again for 99 years. The public fails to recognize that the 100-year flood is really the 1% flood. Specifically, the flood magnitude for the 100-year flood has a 1% chance of occurring in any year. We could experience 100-year floods in two (or more) consecutive years. While this is unlikely, the likelihood does exist and should be prepared for.

Risks that are associated with engineering projects are also difficult to quantify. For example, whether or not a designed structure will experience a class V hurricane over its design life cannot be known in advance. Whether workers who construct a facility will provide quality workmanship or use faulty construction practices is unknown even immediately following the construction. The difficulty in using risk calculations in engineering decision making results from both a lack of empirical evidence and an inability to apply theoretical concepts for elements of a simple system in the risk analysis of complex systems. While considerable theory exists concerning the properties of materials, the safety of a project is imbued with uncertainty when the materials are fabricated, cast, and connected in the final structure and subjected to stress reversals and sometimes the capriciousness of the forces of nature (Gnaedinger et al., 1979). In addition to the risks associated with engineering materials, Gnaedinger et al. (1979) identified several components in construction that contribute to the total project risk; these include the variation of loads that a structure will be subjected to, the behavior of structural elements, the construction process, the ignorance of the designer, the lack of unity in the design/construction team, and the failure of the designer to evaluate all of the risks involved. They further point out that the correlation between laboratory tests and field conditions is fraught with uncertainties.

A cost-benefit analysis is a common way to make risk-based engineering decisions. Risk is often a factor associated with both the benefits and the costs. The potential costs are calculated using the risk estimate and compared to a monetary value estimated for the benefits. If the calculated cost outweighs the benefits, the engineering project or decision should not be undertaken. However, if the benefits outweigh the costs, the decision maker must decide if the difference is sufficient to justify undertaking the project. Therefore, the personal value system of the engineer can become a factor in evaluating a cost-benefit analysis that involves risk. Likewise, the monetary value assigned to consequences such as public safety and fatalities is negotiable. Some may believe that the loss of even a few human lives is not justified regardless of the computed benefits. These uncertainties weaken the credibility of cost-benefit analyses and strengthen the need for good personal value systems among engineers.

CASE STUDY

Consider a county that has 420 unfenced, public storm water management (flood control) structures; on average, the structures have been built over the last five years and are expected to be functional, on average, for another fifteen years. Over the life of the 420 structures, two children have drowned even though "No Trespassing" signs were clearly displayed. Legal settlements of $140,000 and $210,000 were paid in the two cases. The county is considering fencing the facilities, but this is at a cost of $3,000 per facility.

Over the next fifteen years, the expected number of drownings would be six. Thus, the expected benefits would be six times the average of the legal settlements of $175,000; this yields a total benefit of $1.05 million. The cost of fencing all facilities would be 450 ($3,000) = $1.26 million. In this case, the benefit-cost ratio is less than 1.

What decision should be made? What is the risk from the county's perspective? What are the ethical issues involved in the decision?

7.7 PRINCIPLES OF PROBABILITY AND STATISTICS

It would take a book of several volumes to adequately present in detail all principles of probability and statistics that are relevant to engineering risk and uncertainty. The intent here is solely to provide a few basic definitions and a broad overview of the principles. The first important definitions are

- *Random variable*: A quantity that can take any value within a range, but the specific value that will occur at a point in the future cannot be forecasted with certainty—only the chance of the event being a specific value can be stated. Examples: Wind loads on a building, the 100-year flood magnitude, traffic density.
- *Population*: The collection of all values of a random variable that have occurred in the past or will occur in the future; this is rarely known.
- *Data sample*: A collection of events sampled from a population.
- *Probability*: A scale of measurement used to describe the likelihood of occurrence of an event.
- *Distribution*: A mathematical function that describes the likelihood of any value of a random variable occurring at a particular time or location.

Consider the concentration of a pollutant in a river, which is a random variable because it varies in both time and space and a future value cannot be predicted with certainty. We would need to know the concentration at all times in the past and all future values in order to know the population; this would be impossible. A set of ten grab samples would constitute a data sample. Many samples would be necessary to identify the probability distribution of the concentration, but if samples were used to estimate the distribution, then probability statements could be made about the likelihood of future concentration in the river, such as the probability that the concentration will exceed the level that would cause a fish kill. But such an estimate is uncertain, as it has assumed that characteristics of the sample represented the corresponding population characteristics, and that the assumed population is the true population.

7.8 RISK, TECHNOLOGY, AND PUBLIC POLICY: A TRIAD

While they may be defined separately, engineering risk, technological growth, and public policy should be viewed as a triad because of the important implications of the relationships among them. Engineering risk is a link between technological growth and public policy making. While the researcher is involved in much of the basic research that underlies technological growth, it is the practicing engineer who makes it a reality. Therefore, the engineer has a special responsibility to assess the impacts of technological growth. However, it is not widely recognized that public policy plays a major role in the control of risk assessment. The engineer will make a major contribution to public policy making through his or her assessment of risk. Risk assessment is a technically complex aspect of design, so it is an avenue by which engineers can become more influential in public policy making.

Risk assessment is an artform; it is not a completely objective activity. In many cases, the uncertainty in a risk calculation is substantial. Policy makers use risk assessments as input to public policy development, and they use relevant values and political considerations along with the risk assessments to make their decisions. The engineer's assessment of risk is therefore not the only factor used to develop public policy. If the uncertainties in the risk statements are great, then political considerations may be given a greater weight in the decision. Therefore, engineers need to expend the effort to reduce the uncertainty in their assessments of risk. They also need to place considerable effort into risk assessment.

CASE STUDY

Levee failures during Hurricane Katrina were a major issue in terms of the resulting damage. The levees could have been built higher, and then they may not have failed. The hurricane conditions experienced were not expected when the levees were designed and built. Would the additional cost of constructing higher levees be a wise use of public funds? Since the likelihood of such extreme conditions occurring was very low, risk calculations probably indicated that the money could have been better spent on providing protection at other sites. Public policies indicate that public monies be spent on high-risk projects, not for very unlikely events. Hindsight, often said to be 20/20, would now suggest public policy failed the residents of New Orleans.

7.8.1 ETHICS AND RISK

Because of the exponential increase in technological growth and the increase in the options that are created by technological change, the concept of risk assessment is receiving increased attention. Risk has been used for some time as a weighting function in cost-benefit analyses and engineering decision making; however, its corresponding role as a societal value-weighting function has not received similar recognition. Furthermore, the ethical issues related to risk assessment have been virtually ignored. Ethical issues relating to technological growth and risk assessment include the engineers' responsibility to

- Use their specialized knowledge and skill in making accurate estimates of risk
- Provide an estimate of the precision of an estimated risk, as well as the estimated risk itself
- Communicate with laypersons and other professionals the role of risk in decision making
- Understand and respect the roles of other professions and politicians in risk assessment and value conflict settlement
- Maintain competence so that risks can be properly evaluated as technological growth increases societal options
- Affect technological assessment and not just technological growth

The risk-technology-public policy triad can be viewed in many ways. It will be examined herein from the following four viewpoints:

- The role of the engineer: Technical considerations
- The role of the engineer: Human value considerations
- The role of society
- The role of engineering education

7.8.2 THE ROLE OF THE ENGINEER: TECHNICAL CONSIDERATIONS

Engineering risk is used as a decision-making tool in engineering analyses. It is often used in weighting benefits against costs. With respect to technological growth, risk assessment can be an important element in the cost analysis of engineering design, as designs are modified to minimize the adverse effects of technological growth. When design risks are not evaluated, the assessments of benefits and costs may be inaccurate, which can lead to poor decisions. While we can put a value on the loss of a building, assessing the value of the loss of a life is more difficult. Assessing the worth of losing an endangered species would be quite subjective. Societal costs associated with risk can be very significant. Engineers have an obligation to provide the technical expertise to those involved in formulating public policies that will minimize risk losses due to the uncertainties in engineering design. Estimates of risk are also uncertain. Quite often, too little effort is placed on uncertainty assessment; this may lead to both inadequate public policy and uncontrolled growth.

It is difficult for risk assessments to take place before technological change occurs. However, the engineer has an obligation to use data that resulted from growth related to past technological change to increase the accuracy of risk assessments. Knowledge of the precision of an estimate of risk is just as important as the estimate of risk itself. Quite frequently, the uncertainty in engineering risk assessments creates public concern about technological growth; the uncertainty makes it difficult to resolve value conflicts as the competing parties both have inaccurate estimates of risk.

Public policy has a significant effect on both risk assessment and technological growth (Tribus, 1976). Fear of the unknown can skew the development of public policies. The engineer must help public officials identify both how public monies can be best allocated for assessing risk of design factors and the design methodologies that will maximize technological growth within the bounds of public policy. The former generally requires research. Unless the engineer actively participates in the formulation of public policy, the balance between funds spent for technological growth and research on risk assessment may not be optimum; furthermore, poorly formulated public policies may place excessive control on technological growth.

7.8.3 THE ROLE OF THE ENGINEER: HUMAN VALUE CONSIDERATIONS

In addition to technical considerations, the engineer has a major role in human value considerations among risk, technological growth, and public policy. Assessments of risk can be used to weight conflicting human values that arise as part of technological growth. When the engineer reduces the uncertainty in risk, conflicts between human values will become less of an issue. Additionally, the engineer must place the knowledge in a form that can be understood by those who do not have the special knowledge and skills of the engineer. Then, the public can participate in solving the ethical dilemma through democratic institutions. That is, the engineer must develop the methodologies for using his or her risk assessments to help politicians formulate public policies that best reflect societal values.

Technological growth is intimately related to human values. In many cases, public policies and risk assessments must be reformulated after new risks and value conflicts that result from new technologies are identified. The engineer has a responsibility to ensure that rational policies result; irrational policies may result when the reformulation of a policy results from emotional analysis of poorly understood technological growth.

7.8.4 THE ROLE OF SOCIETY

While the engineer plays an important role in the triad, society has an obligation and must remain responsible for the effects of technological growth. First, society has an obligation to be knowledgeable about the risks that accompany technological growth. Quite often, society forms opinions about engineering risks using information obtained from special interest groups rather than estimates of risk obtained from proper engineering analyses (Florman, 1976). Second, society must elect public officials who will make engineering risk an integral part of public policy formulation; the public must accept the responsibility for technological growth that results from public policies. Third, society has a responsibility to provide an atmosphere where technology can grow and engineering risk is properly assessed; therefore, they must provide public funds and encouragement for research into risk assessment.

7.8.5 THE ROLE OF ENGINEERING EDUCATION

In part, engineers are a product of their formal education; thus, it is of interest to examine the role that education should play in preparing engineering students for participating in risk assessments that serve as inputs to public policy formation. A significant part of engineering education involves the introduction of design methods. Quality education includes instruction about methods of evaluating the effects of design uncertainties. While basic data analysis techniques are introduced, the instruction often stops short of material on risk assessment.

The topic of professionalism also does not receive the attention it should in engineering education. A course in professionalism should be a part of every engineering curriculum and aspects of public policy, and its role in engineering should be introduced. Specific aspects of public policy making that are of special interest to the engineering student are the institutional framework for the engineer to provide input to policy making and the role of public expenditures on the advancement of state-of-the-art engineering. The relationship between technological growth and public policy should also be introduced; of special importance is the potential for public policy to both advance and limit technological growth. Such topics would help prepare the student for an active role in the relationship among technological growth, public policy, and risk assessment.

7.9 RISK AND VALUE CONFLICTS

From an engineering viewpoint, risk is defined as the product of the probability of an event and its consequences. From an economic viewpoint, risk is associated with the potential for damage. Certainly, the politician views risk more pragmatically. While all of these viewpoints are important, the concept of risk must be viewed in a broader context if the societal value of technological growth is to be optimized.

Value conflicts, which arise when one value is in opposition to another, are often an outgrowth of technological change. If the engineering risks associated with the technological changes could be quantitatively evaluated precisely, then the best solution to any ethical dilemma would be apparent. The difficulty in solving the ethical dilemma increases as it becomes more difficult to accurately evaluate the risks. It becomes apparent, then, that engineering risk acts as a societal value weighting function. That is, risk is a primary factor that individuals, or groups, use to weight the societal values that are in conflict. Any disparity in the assessment of risks, whether it is by the engineer or the politician, will be translated into variation in the weights assigned to opposing societal values. That is, value conflicts result from inaccuracies in the estimation of engineering risks.

A little reflection will make it evident that risk as a societal value weight is highly correlated with the more familiar probabilistic interpretation of risk, which is also used as a weight for economic alternatives. In its usual applied definition, risk reflects uncertainty in economic decision making. With respect to decision making in a value orientation, risk reflects the fact that all fundamental values are important, but that individuals are uncertain on the weight that should be given to each fundamental value when these values are in conflict.

Engineers play a vital role in technological growth. While public needs create a situation in which value conflicts are inevitable, assessments by engineers advance societal values. Technological growth increases the available technical and societal value options, while engineering risk is used to weight the damage potential to both the physical state and the value state of society. The major problem that engineers are confronted with when using risk as a value weighting function is the specification of relationship between risk and values.

While the engineer is a vital link in the relationship between risk assessment and technological growth, the meaning is complete only when their interrelationships with public policy are considered. Properly formulated public policies can minimize both institutional conflicts and controls and maximize the benefits of technological growth. Poorly formulated public policies can increase the risks of both technological failures and societal value conflicts. The engineer is a central figure in ensuring that public policy and technological growth are properly carried out with minimal risk to society.

7.10   DISCUSSION QUESTIONS

1. Tables of Manning's roughness coefficient (n) in fluid mechanics books often give a range of values for each type of surface. For example, the n for concrete may indicate a range from 0.011 to 0.017. Discuss reasons for the uncertainty in n.
2. Identify uncertainties in knowledge of global warming.
3. Identify uncertainties in knowledge of hypoxia (see *Civil Engineering*, June 2008, p. 54).
4. Identify the inputs to the design of a simple cantilevered beam and assess the uncertainties in both the inputs to the design and the construction of such a structure.
5. Obtain a table of runoff coefficients of the rational equation for computing peak discharge rates. Discuss factors that contribute to the uncertainty in values and lead to a recommended range of values.
6. Steel corrosion depends on many factors. Identify factors related to knowledge that influence the corrosion rate and discuss the uncertainty of each factor. Also identify factors that contribute to uncertainty in data collected to develop a prediction equation of corrosion loss with time.
7. Define the Atterberg limits for soils and discuss their uncertainty in knowledge and in values of the limits estimated through laboratory experiments.
8. Flood frequency curves, which relate the magnitude of a flood and the probability that a flood of a specific magnitude will be equal or exceeded, are uncertain because of a lack of adequate data and the inability to identify the

appropriate theoretical probability distribution function. Discuss sources of database and knowledge-based uncertainties.

9. Estimates of evaporation of water from lakes are not precise values. Measured data can improve the accuracy of computed estimates, but the theory that underlies evaporation is subject to uncertainty. Explain the sources of these uncertainties.

10. Theoretical considerations can be used to develop a strength envelope for a given soil. However, an empirical method such as Coulomb's law is often used. Discuss uncertainties in theory, and with an empirical method such as Coulomb's law.

11. The simple t test on the mean of a sample of engineering data is used to test hypotheses. Discuss the implications of using the t test because of the uncertainty in sample measurements.

12. Confidence intervals are a statistical method that indicate uncertainty in a statistic, such as a mean or a regression coefficient. If a regression model were developed between steel corrosion and time and a confidence interval was computed on the slope coefficient, what would the confidence interval represent in terms of uncertainty in steel corrosion estimates?

13. Five surveyors measure the difference in distance between two points, with each estimate differing from the other four. What factors contribute to the uncertainty in the values? What value would you use as the best estimate, and how would you indicate the accuracy of your estimate?

14. In the case study of the benefits and costs of fencing the storm water facilities, what decision would you make? Justify your decision. If someone opted to order the fences to be installed, is the difference $0.21 million that cannot be spent on other public works? How can this be justified?

15. Person A is an avid skier and takes many risk, such as skiing in areas where avalanches are quite possible; however, the person gave up smoking because of the potential risk of getting cancer forty years later. Person B, a friend of person A, smokes but refuses to go skiing because of the risk. Discuss the rationality of A and B's risk decision making.

7.11 GROUP ACTIVITIES

1. Concrete is a primary building material; however, it is a composite of other materials. Identify the materials used to make concrete and identify factors that can contribute to the uncertainty in the compression strength of concrete.

2. The earthquake intensity at a site in an earthquake-prone area is quite uncertain. Identify factors that contribute to the uncertainty of a computed intensity at a specific site where the construction of a high-rise building is being considered.

3. In July 1978, President Carter directed the U.S. Water Resources Council to assess ways of including risk and uncertainty in federal water projects. Obtain a copy of the directive and comment on the role of risk analysis in federal projects.

REFERENCES

Florman, S. C. 1976. *The existential pleasures of engineering*. New York: St. Martin's Press.

Gnaedinger, J. P., Hanson, W. E., and Khachaturian, N. 1979. Risks and liability in engineering practice. *Issues in Engineering, Journal of Professional Practice*, ASCE 105(EI4):207–17.

Lee, L. T., and Collins, J. D. 1977. Engineering risk management for structures. *Journal of the Structures Division*, ASCE 103(ST9):1739–56.

Siekevitz, P. 1970. Scientific responsibility. *Nature* 227:1739–56.

Tribus, M. 1976. Along the corridors of power—Where are the engineers? *Mechanical Engineering* 57(4):24–28.

8 Communication

CHAPTER OBJECTIVES

- Identify characteristics of effective communication
- Discuss organizing communication responsibilities
- Provide guidelines for developing communication skills
- Discuss communication with nontechnical audiences
- Provide guidelines for evaluating professional communications

---

8.1 INTRODUCTION

Studies have reported that communication skills are more important to engineers than the technical skills that form the core of most civil engineering undergraduate programs. Yet in spite of the importance of communication skills, engineering graduates are ranked as poor communicators. Many reasons could be cited for this deficiency, but the reasons are not important. Ways of correcting the deficiency are important. First and foremost, the civil engineering student must recognize the importance of communication and accept the fact that he or she can greatly benefit by becoming a better communicator.

Entire books are devoted to the enhancement of communication skills. This chapter will not attempt to duplicate the content of these books. Instead, emphasis will be placed on characteristics of effective communication and some of the problems that lead to ineffective communication. Both undergraduate and young professionals should make every effort to take courses that include writing and speaking activities and read books that will lead to improved communication. This is an important part of lifelong learning.

8.2 VALUES RELEVANT TO COMMUNICATION

Communication in engineering has a value basis. Ultimately, a poorly communicated design could lead to failure that negatively affects the public. More likely, poorly communicated work reduces the reader's efficiency. Thus, poor communication skills can hinder success. The following are values that are relevant to communication:

- *Knowledge*: Understanding gained through reading or listening to someone else's experiences.
- *Efficiency*: Producing an effective communication with no unnecessary waste of effort by the writer or reader.

- *Promptness*: Producing the communication in a timely manner.
- *Care*: Being attentive to detail.

Communications are generally attempts to transmit knowledge. Therefore, a communication should be well prepared so that the intended message is efficiently and accurately transmitted. Efficiency is important to both the writer and the reader, as the loss of time means that other responsibilities may not be fully met. Care must be taken in preparing the communication so that ideas and findings are accurately stated in the communication. Promptness is important, as lateness often causes people to have inadequate time to digest the material once they receive it. Obviously, procrastination must be avoided, as it frequently leads to poor communication.

8.3 CHARACTERISTICS OF EFFECTIVE COMMUNICATION

Given that engineers may spend as much as 50% of their workday either writing reports for others to read or reading other peoples' reports, the effectiveness of the writing is clearly important. At a minimum, a poorly written document reduces a reader's efficiency, as he or she will probably need to reread many parts and then spend time deciding about the points that the writer was trying to make. Writing effectively is not easy. Understanding the factors that contribute to effective writing is the first step in improving writing skills. The following is a list of factors important to writing well and improving one's own writing:

- *Active construction*: Avoid passive voice.
- *Clarity*: The meaning of all statements needs to be unambiguous.
- *Definitive*: The prose should be precise, not vague.
- *Repetition*: Ideas and conclusions should not be repeated.
- *Confidence of statements*: Each sentence or group of sentences should sound confident rather than weak.
- *Economy of words*: Sentences should not be overly wordy, as the meaning of wordy sentences is often unclear.
- *Variety*: Sentences should be varied in length and structure to ensure readability.
- *Language appropriateness*: The sentences should be appropriate for the intended audience.
- *Emphasis*: Words should be selected to provide the proper emphasis to the meaning.
- *Overuse*: Clichés and hackneyed words should not be used.
- *Orderliness*: The sentences within a paragraph and the paragraphs within a section should be logically ordered.
- *Proportion*: The ideas/topics should be given space in proportion to their importance to the reader.
- *Thematic*: Each paragraph should center on a common topic, and paragraphs within a section must be related.

- *Insight*: Each sentence or paragraph should communicate important information.
- *Accuracy*: Every conclusion should be appropriately supported.
- *Credibility*: All statements should be believable, with debatable statements backed up with clarifying support.
- *Support*: Generalizations should be supported, with sufficient detail provided such that the reader would agree with the generalization.

These characteristics were composed with reference to written communication, but they apply to other forms of communication as well.

8.4   DEVELOPING GOOD COMMUNICATION SKILLS

What are people's greatest fears: snakes, spiders, and speaking? Most people dread standing before a group and presenting their ideas. Part of the fear often stems from a bad experience. For example, a person may have been asked to act as a stand-in for someone who did not show up, and because the person was unprepared, the presentation went poorly, to the extent that a fear of speaking developed. Just as a fear of speaking develops, the fear can be conquered, but this requires an attitude that supports the belief that being a good communicator is important. The person must also have the confidence to master the skill.

The following four phases constitute the ROPE method of overcoming the fear of oral communication: read, observe, practice, experience. The first action recommended for becoming a good speaker is to read a few books on making speeches. Each of these books provides guidelines such as "know your audience" and "have good visual aids." The primary benefit of reading these self-help books about communication is that they will help you develop a plan of action, and they generally give case studies of other people who have overcome their fear of speaking. When reading such books, it is a good idea to make lists of do's and don'ts—things to do when making a speech and things to avoid doing. These lists can then be reviewed when preparing a presentation.

The second phase of the process involves observation. Attend presentations by professionals or those who you would expect to be good speakers. Observe how they do it, taking special note of their actions that you believe add positively to a presentation and those actions that seem to detract from speaking effectively. For example, did the speaker immediately engage his or her audience with the opening remarks? This would be a positive. Did he or she spend more time looking at the screen that showed his or her slides or directly at the audience, making eye contact? Looking at the screen is considered a negative. Then when you are preparing for a presentation, consider both the positive and negative actions to improve your chances of success.

The third phase of the process emphasizes practice. First, get a small group of friends who have a similar fear of speaking and have informal meetings where everyone must prepare a short, maybe three- to five-minute, presentation on any subject. In addition to the prepared topics, each person should be given a topic and asked to extemporaneously speak on the topic. A person could, for example, be asked to talk

FIGURE 8.1   The Feelings–Action Loop.

about leadership, his or her recent summer job or internship, or the value of graduate school. These impromptu presentations are intended to provide an experience of organizing and planning a talk on short notice. It will also help in learning how to respond to questions that follow a presentation. Just before making a formal presentation, practice is extremely important.

Finally, gaining experience is necessary to overcome the fear of presenting before a group. Experience can be gained by volunteering to be the presenter of group activity reports; taking courses, especially those in the humanities or social sciences, where speaking opportunities are part of the course requirements; and becoming an officer in a student group where reporting on activities or leading a meeting is required. The latter activity will provide other leadership benefits.

Overcoming the fear of speaking is not easy, but following the steps of the ROPE process will likely erase the fear. Remember the feelings-action loop, shown in Figure 8.1. Bad feelings engender bad action. Then the bad action engenders bad feelings in the future, which then keeps the speaker in a negative loop. Following the ROPE process will use this same loop, but in a positive way. Specifically, positive feelings based on reading, observation, and practice will engender positive experiences, which will lead to good feelings about future presentation responsibilities.

While these steps may eliminate the fear of speaking, we still need a plan for overcoming the fears of snakes and spiders.

8.5   PRESENTING TECHNICAL MATERIAL TO A NONTECHNICAL AUDIENCE

Presenting technical material to your peers may seem difficult, but in reality, it is much easier than presenting technical material to a nontechnical audience, especially when the topic places you in an adversarial position with the audience. For example, if your position is to explain to a community group why the best location for a new waste treatment plant is just a half mile from their community, you will likely be facing an ill-disposed, hostile group. Your task will be especially difficult if the individuals in the group have developed many misperceptions about the project. Even in cases where a group is there only to learn and they are not in conflict to your position, summarizing technical material can be difficult. Following a few guidelines can improve your chances of having a successful communication with the nontechnical audience. The guidelines will be separated into three groups: preparation, communication, and follow-up.

8.5.1 PHASE I: PREPARATION

The following guidelines relate to the communication process with a nontechnical audience; considering these items can make the actual encounter more effective:

- Listen to their description of the problem while paying special attention to their misperceptions and level of emotional involvement.
- Identify the real problem and break the problem into fundamental elements that can be addressed individually.
- Learn exactly what they believe will represent an amicable solution to the problem, i.e., know what they want to accomplish.
- Identify and isolate each of their misperceptions so that each can be addressed separately.
- Ensure that they recognize the fault of each misperception before presenting new concepts to develop correct perceptions.

Once you understand the problem as they see it, know what they want to accomplish, and have overcome their misperceptions, you can present material to help them understand your points of view.

8.5.2 PHASE II: COMMUNICATION

Success at this phase of the communication will require honesty, style, and experience, not just a strong technical background. Guidelines relevant to the technical communication are:

- Place the problem in context by providing background information and the basics of the underlying physical processes; this is the time for knowledge transmission.
- Work to get the audience to see the problem from an engineering perspective.
- Be honest about any uncertainties related to the issues that you present.
- If they have trouble with technical issues, try to develop analogies or hypothetical situations that have similarities with their problem.
- Minimize the use of technical jargon and uncommon acronyms.
- Do not go beyond your level of expertise.
- Stick to the facts and avoid personal biases.

Style is important, as it can communicate a sense of leadership, sincerity, and competency to the audience. Relevant guidelines related to style are:

- Maintain a professional demeanor when presenting technical concepts and when responding to questions.
- While you should not provide technical details that the audience cannot understand, you should also not talk below their level of knowledge.
- Show confidence when presenting material or when responding to questions.
- Control the speed and volume of your delivery.
- Be conscious of body language.

Your method of delivery is also important. Guidelines relevant to the format of your presentation include

- Use graphics and mental images to illustrate technical concepts.
- Periodically summarize important points and points of agreement to ensure that the audience understands their significance.
- Use lists to summarize conclusions.

8.5.3 PHASE III: FOLLOW-UP

Following up after a meeting is generally a positive action, as it shows concern for the issues discussed at the meeting. Guidelines relevant to follow-up activities include

- In a memorandum, summarize conclusions, and both agreements and remaining disagreements.
- If you or they agree to do follow-up work, such as collect additional data, make observations, or provide new analyses, set a timeline for the completion of this work.

8.6 GENERAL STRUCTURE OF A REPORT

The structure of a report will depend on the nature of the project and the intended audience. However, the following components are likely to be part of any report:

- Executive summary or abstract
- Introduction
- Methodology used to solve the problem
- Analyses made
- Results
- Conclusions
- References
- Appendices

Appendix A provides some basic observations about each of these components.

8.7 GUIDELINES FOR ORAL PRESENTATIONS

The oral communication process can be separated into four steps: (1) formulating the presentation, (2) compiling the material for the presentation, (3) rehearsing, and (4) making the presentation. It is important to recognize that a poor presentation almost always results from failure during the first three steps, not the fourth step. If sufficient attention is given to the preparatory steps, then the actual presentation will most likely be successful. Proper attention to the first three steps can also help reduce nervousness, which is usually the number one concern of the novice.

8.7.1 Formulating the Presentation

The best way to formulate a presentation is to answer the question "What major point(s) should be made?" By focusing on the major conclusions of the presentation, one can then prepare to educate the audience.

- *Know your audience.* A speech to be presented to a homogeneous audience will be different from one prepared for a heterogeneous audience.
- *Educate your audience.* In preparing the speech, identify the educational objectives of your presentation. What new knowledge do you want the audience to have after listening to your speech?
- *Entertain your audience.* The audience is more likely to grasp the educational points of your presentation if it is presented in an entertaining fashion. Effective visuals provide the opportunity for making a presentation more entertaining, but make sure that the approach to entertainment will be well received by the audience and not viewed as sophomoric.
- *Persuade your audience.* Your presentation should include strong supporting material in order to persuade the audience to believe in your conclusions.
- *Be mindful of time constraints.* When formulating a presentation, know the time allotted to the speech and plan only to develop material that can be effectively presented in the allotted time.

8.7.2 Developing the Presentation

A very efficient way of developing a presentation is to use the progressive outline approach. With this method, a very simple outline of four to six lines is made to address the following questions:

1. Why was the work done? (State problem and goal.)
2. How was the work done? (State solution method.)
3. What findings resulted from the work? (State one or two major conclusions.)
4. What do the results imply? (State the implications of the work.)

Guidelines related to these questions are as follows:

- *Content of subsequent outlines.* With each outline, add more content that relates to the educational objectives identified in the speech formulation phase.
- *Begin organizing visuals.* Visuals add variety to a presentation and can serve as the medium for entertainment and education. Visuals outline the presentation for the audience and can serve as cues for the speaker.
- *Give special attention to the introduction.* Nervousness is most severe at the beginning of a speech, so a well-developed introduction can increase the speaker's confidence. Also, a poor introduction will cause the audience to reduce its attentiveness.

- *Focus on the conclusions.* The conclusions to a speech are important because the end of the speech is when the speaker identifies the major points. The conclusions are the points that the audience should learn from the speech.
- *Capture the audience's attention.* Make sure the major points are covered without trying to do too much. Presentations that are crowded by too many details are often ineffective.

8.7.3   REHEARSING THE PRESENTATION

Rehearsing is important to reduce nervousness, to ensure that the time constraint will be met, and to become sufficiently familiar with the material that notes will not be needed. A few guidelines related to rehearsal are as follows:

- *Location, location, location.* Try to practice in the exact place where the speech will be given. Familiarity with the surroundings helps put a speaker at ease.
- *Practice the opening statements.* Nervousness is greatest at the start of a speech, so be extra familiar with the opening remarks to ensure that these capture the attention of the audience.
- *Have friends critique the presentation.* Rehearsing with an audience rather than by yourself will make the rehearsal more like the actual presentation. The friends should also be willing to give serious criticism so changes can be made before the presentation.
- *Do not rehearse in silence.* When rehearsing, speak aloud. If you just rehearse by mouthing the words, you will not be able to judge the time because you speak slower when speaking aloud than when rehearsing in silence.

8.7.4   MAKING THE PRESENTATION

If you were successful with the first three phases of the oral communication process, then chances are that the actual presentation will be successful. A few guidelines relative to making the presentation are:

- *Nervousness is good.* Some nervousness is to be expected and can be beneficial if it makes you concentrate more on your presentation and less on the audience.
- *Make eye contact.* Making eye contact with those in the audience will help keep them engrossed in your presentation; however, you do not want to think about the person with whom you make eye contact.
- *Avoid filler words.* Words like *uhm, uh,* or *you know* are called filler words, as they are spoken, often unbeknownst to the speaker, to fill the time gap between ideas. Short gaps of silence are not necessarily bad, as they give those in the audience time to think about what you have said. However, continued use of filler words can be distracting to the audience.

- *Be careful of bad body language.* Your hands can be distracting to the audience if you motion too much or not enough. Also, do not fold your arms in front of you, as this indicates that you are closed to the audience. Keep your hands out of your pockets as well. Good body language can help your presentation succeed.

8.7.5 RESPONDING TO QUESTIONS

The question-and-answer period in very important, and questions should be encouraged. Questions show that you have sparked their interest. When you are asked a question, do not be afraid to pause for a moment to formulate a response. Try to keep your response short and to the point. If you are asked a question and have no idea about how to respond, ask the person to rephrase or clarify the question. If you understand the question but do not know the answer, be honest and state that you do not know. If a person seems combative when asking the question, do not try to match his or her attitude; it is better to respond in a very neutral tone and move on to another questioner.

8.8 LISTENING AS A COMMUNICATION SKILL

True or false: Engineering managers spend as much as 50% of their on-the-job communication time listening? This is actually true, as managers spend a considerable amount of time in meetings, including those with clients, and in having subordinates explain their progress.

True or false: Emotions are one of the most significant barriers to listening? Again, this is true. An emotional response to a point made by the speaker can prevent a person from receiving the intended message. Emotions can cause the listener to be biased, even defensive.

True or false: Listening skills become more important as a person takes on more leadership responsibilities? Very true! A leader depends on others for information, whether it is a potential client explaining the task, subordinates providing a summary of their work results, or members of the community criticizing a proposed project. Poor listening skills prevent the transfer of knowledge, which can cause a person in a position of leadership to make poor decisions.

True or false: Listening is an innate skill so a person cannot become a better listener? This one is false. Listening can be learned. A few ideas on becoming a better listener are as follows:

1. Thought is faster than speaking, so in your mind, try to summarize what the speaker is saying as you listen.
2. Do not immediately start to think about a reply while the person is speaking, as you will then be concentrating on your reply rather than the speaker's message.
3. Avoid being distracted by the speaker's mannerisms or events taking place around you.

4. Think about the topic prior to the meeting and identify ideas about which you would benefit from knowing.
5. Try to find positives in the message, not just negatives.
6. If it is practical, take notes, especially those that identify the speaker's key points.
7. If the speaker takes an extended period of time to present his or her argument, ask him or her to summarize key points.

Key problems that result in poor listening are

1. Based on your knowledge of the speaker, forming a negative opinion of the speaker's points before he or she has presented them.
2. Do not fake attention to the speaker, as this indicates that you are not listening.
3. Do not tolerate distractions from other people who are around you and the speaker. Move to a quiet area so that you can concentrate on the speaker.
4. Do not get mentally sidetracked by difficult material. Have the speaker slow down or repeat key points.

Listening and hearing are not the same, as hearing is passive, while listening is active. Listening requires interpretation of sounds. In many cases, this involves filtering out unwanted sounds so that relevant information is processed. The listener makes an active effort to extract useful information.

Leaders tend to be exceptional listeners. They practice critical listening, which involves both comprehension and evaluation of a message. This increases their efficiency of information gathering as well as their understanding of the message. Critical listening requires the listener to actively think while the speaker is talking. The critical thinker overcomes the barriers to listening.

8.9   SELF-EVALUATION IN COMMUNICATION

Educators talk about two types of evaluation, formative and summative. Very briefly, formative evaluation is assessment *for* learning, and summative evaluation is assessment *of* learning. Formative evaluation takes place prior to the product being complete, with the purpose of improving the product. In the case of a written communication (e.g., technical paper or report), formative evaluation can be conducted on drafts, with the assessments intended to find weaknesses that can be corrected prior to submission of the final version. Hopefully, formative learning will accompany the formative evaluation.

A summative evaluation is made after the product is completed and submitted. In terms of a classroom writing project, the summative evaluation would lead to a grade and, hopefully, detailed comments on the criteria used in evaluating the document. In cases where the product had the intent of solving some problem, such as the cleanup of a hazardous waste site, a summative evaluation may assess the extent to which recommendations made in the report solved the problem. Such a report might also be summatively evaluated on the extent to which the report serves as a guide for reports on similar problems at other locations. These products do not receive letter grades, such as classroom paper assignments receive, but are generally qualitatively assessed based on a success-to-failure scale. The intent of the summative evaluation is not learning, although that may occur following the evaluation.

The primary emphasis here is on a formative self-evaluation of a writing effort. Self-evaluation is difficult but extremely important. While evaluating someone else's written document may be easier than evaluating your own work, using a systematic formative evaluation of your own work can lead to both a better product and formative learning. While a diagnostic evaluation of your own writing may be the immediate objective, identifying your own attitudes and biases about writing responsibilities may be the greatest benefit of formative self-evaluation.

Formative self-evaluation of a written draft can focus on both the grammar/organization of the draft, hereafter referred to as rhetorical evaluation, and the technical content. The guidelines for self-evaluation herein focus on rhetorical criteria.

If a written document is to be successful, it must continually undergo a formative assessment. Ideally, this would be done by an independent evaluator, but in many cases, the writer will need to take on the role of the evaluator. A self-evaluation has the greatest chance for success if the writer uses a set of evaluation guidelines and sets aside a time when he or she acknowledges that evaluation of the most current draft is the objective.

A systematic self-evaluation should be conducted periodically, maybe even frequently. Each evaluation could have a specific objective. The following are types of evaluation that can lead to improvement of the document:

- *Needs evaluation*: The writer reviews the manuscript to ensure that each part of the manuscript is relevant to the needs of the person or group for whom the manuscript is being written.
- *Objectives evaluation*: Ultimately, the document will not receive a good summative assessment if the objectives behind the effort are not met. Therefore, the following question should be continually asked: Is what I am writing directly relevant to the objectives?
- *Process evaluation*: The focus of this type of evaluation is the way that the material is being delivered. The evaluator should assess if material would be better presented as prose, tabular data, or graphical summaries.
- *Rhetorical evaluation*: Rhetorical evaluations are broad, as they include the assessment of the grammar, style, organization, and strategy of the writing. Therefore, when performing a rhetorical self-evaluation, the writer should consider the following factors:
  - Correctness: Is the punctuation correct? Are references properly formatted?
  - Organization: Is the sequencing of ideas logical? Is the need discussed before the goal/objectives? The experimental design discussed before the analysis? The findings presented before the implications?
  - Style: Is the wording/vocabulary appropriate for the intended audience? Will the reader find the material bland, or does it provide variety in tone?
- Strategy: Did the writing effort follow a systematic practice?

In general, evaluations are made to provide feedback about the merits of some effort. For example, a teacher evaluates the work of a student and then provides a grade to reflect the worth of the work. The evaluation discussed in this section is aimed at the reader to use in self-evaluation of communications. A reader should make continual evaluation criteria prior to preparing a communication, oral or written; a person can improve the work before submitting it for a summative evaluation by someone else.

8.10 DISCUSSION QUESTIONS

1. Why do people fear public speaking? List reasons.
2. Explain why the characteristics of clarity, definitiveness, and repetition are important when writing an engineering report.
3. Explain why economy of words, variety, and emphasis are important when writing an engineering report.
4. Why should each paragraph include a topic sentence?
5. Why is variety in sentence length and structure important?
6. List reasons why outlining the content of an engineering report prior to putting the material in prose format is beneficial.
7. Develop a set of rules for constructing an outline.
8. What are the benefits of proofreading your own writings?
9. Why are communication skills especially important when presenting material to a nontechnical audience?
10. How would word selection for a report to a nontechnical audience differ from that for a technical audience?
11. Discuss ways that an engineering report intended for a nontechnical audience would differ from a report written for a technical group.
12. Identify ways that a person can become a better speaker.
13. Why does a leader benefit from being a good listener?
14. Using a previously written laboratory report or term paper, critique it based on the concepts discussed in this chapter. Make a list of ways that it could be more effectively written.
15. Develop a set of rules for presenting engineering data in a histogram format.
16. Develop a set of rules for presenting engineering data in an x–y plot.
17. Explain the benefits of graphically presenting engineering data.
18. Develop a set of rules for constructing a table to present engineering data.

8.11 GROUP ACTIVITIES

1. Obtain one civil engineering report, one government document, and one article from a civil engineering research journal. Compare and contrast the following: (1) the format used for headings, (2) the average length of headings, (3) the format used for references, (4) the quality of the figures, and (5) the way that data are summarized in tables, including the format of the column headings.

2. Obtain a set of PowerPoint slides that were prepared for a technical presentation on an engineering topic. This could be from a professor's lecture, but not one for which you were in attendance. Fully evaluate the slides and make a list of positive elements of the slides, and then a second list of negative points that should be changed. Discuss the extent to which the slides by themselves enable the viewer to learn about the topic.
3. Obtain a copy of a student's brief report, about five double-spaced pages. Read and critique the paper for structure, grammar, and effectiveness. Evaluate the use of headings, paragraph structure, and topic sentences, as well as the graphics.

9 Public Policy

CHAPTER OBJECTIVES

- The principal elements of public policies are identified
- Ways of participating in the development of public policies are discussed
- An analysis of a public policy is made
- Ethical issues relevant to public policies are discussed

9.1 INTRODUCTION

From a management standpoint, a *policy* is a plan that states in very general terms actions that are needed to meet organizational goals. This definition can be modified to reflect policies developed by governments, which would be referred to as public policies. These policies address goals considered actions important to society. Public policy is of interest to the engineering profession because of the societal values that are embedded within public policies and the effect of policy on engineering design and practice.

Technological growth introduces new options to both an individual and society; for example, the debate over the benefits and risks of nuclear power would not exist if technological growth had not introduced nuclear power as an option. The options often create value conflicts between special interest sectors of a society. Engineers are responsible for a considerable portion of technological growth. Engineering has been subjected to a significant level of criticism because technological growth creates conflicts, sometimes introducing risks, real or perceived. Many of the concerns about new technologies may be warranted. The engineering profession has a responsibility to society to identify and help minimize the risks associated with technological growth; this will minimize value conflicts. Engineering risk is the link between technological growth and societal values, and thus is important in assessing and solving societal value conflicts.

Civil engineers have considerable technical knowledge that is often wasted because of their lack of appreciation for public processes and their difficulty in dealing with the necessary compromises that accompany the political process. In the late 1970s, I attended a workshop where one of the speakers was a city councilwoman. I have always remembered one of her comments, which went something like: "I cannot use many of the recommendations made by engineers because they are unwilling to recognize the political compromises that I have to make to get reelected." As many issues such as energy and the environment become more important to society, the civil engineering profession that is uniquely qualified to deal with associated techni-cal issues will need to be more sensitive to public policy and the political process. Failure to recognize that knowledge of public policy is important will limit the ability of the profession to meet its technological and ethical responsibilities to society."

9.2   PUBLIC POLICY

Policy consists of the following three parts, which collectively reflect the public intentions of the policy and the products of its implementation: (1) the goals, objec-tives, or commitments of a political unit; (2) the means selected for implementation or obtaining these goals; and (3) the consequences of the means, i.e., whether, in fact, the goals are actually realized. The values of a society are reflected in its public policy. Engineering policy refers to public policies in which the engineer has signifi-cant influence in their formulation or application.

While civil engineers have a relatively low level of influence on public policies that concern foreign aid, agriculture, and military disarmament, the civil engineer's potential influence on public policy in arenas such as energy, infrastructure, the environment, and transportation should be significantly greater than it currently is. Factors that determine the level of influence by engineers include (1) the degree of conflict about technical matters among engineers, (2) the personality of engineers; (3) the degree to which engineers seek the power needed to influence policy mat-ters, (4) the level of technical complexity of the policy arena, and (5) the presence of vested interests. The importance of these factors will vary with the policy arena. However, engineers need to know how to effectively influence public policy, in the process of both formulating public policies and evaluating public policies, so that they can influence the implementation of new policies.

9.3   TECHNOLOGICAL GROWTH

While providing a comprehensive definition of technology is difficult, the public views technology as the mode by which society is provided with the material goods that are necessary to improve their standard of living. Some view technology in a more narrow perspective; specifically, they view it as the application of scien-tific principles for industrial advancement. In fact, technology should be viewed in a much broader sense; technology is a process viewed as providing benefits to society. That is, technology should be viewed as the methods used to advance all aspects of society, including the state of human values. Engineering design is a systematic, codified process and is central to technological growth.

Engineers have played a major role in many areas of technological growth; how-ever, they have sometimes failed to gain their due recognition for their positive role. In the early 1800s, engineers pushed for policies that would improve the safety of steam boilers, such as those used as the power source for steamboats. It took them almost fifty years to effect change, partly because of their lack of appreciation for the political processes related to developing public policies. Recent discoveries of detrimental side effects of technological growth, especially environmental damage, have spawned criticism of the failure of engineers to properly assess the potentially negative impacts of technological growth. Global warming is the most notable side effect of the population and technological growth of the last century.

The criticism of engineers is partly valid because engineers have failed to actively participate in policy making. Tribus (1978) discussed the need for engineers to more actively participate in policy making. Florman (1976) discussed the stereotyping of the engineering personality and indicated that the typical engineer is generally indifferent to public affairs. He goes on to point out that engineering does not create the policy-making-adverse person; rather, it is that sort of person who is choosing engineering. It would be easy to add that engineering education does little to introduce public policy, and that this perpetuates the problem. However, it shouldn't be this way, and it doesn't have to be that way.

9.4   ACTIVELY PARTICIPATING IN PUBLIC POLICY

A civil engineer can participate in public policy in a number of ways. These range from the simple act of writing a letter to the more difficult act of running for a political office. Numerous options fall in between these extremes, such as meeting with political representatives, communicating in person with officials in government agencies, actively participating with community groups, or serving as an intern or fellow in a congressional office.

9.4.1   LETTERS TO THE EDITOR

A letter to the editor of a newspaper or weekly news magazine is a simple yet effective way of entering the public policy arena. For a local issue, a letter to a local newspaper about a current topic of interest, especially if it relates to a recent story in the paper, can illustrate the value of the technical knowledge possessed by civil engineers. For example, if technical facts in a recent story were wrong or presented in a biased way, a well-composed letter that corrects the facts would be welcomed by the paper. If a national issue has local implications that have not been addressed in an article in the local newspaper, then publishing a letter to the editor may be a useful way for the paper to make the readers aware of the local effects. The engineer's technical knowledge is a valuable resource for the public.

The chances for having a letter published increase if a few general rules are followed:

- The technical issues should be written in a way that the typical reader would understand. The effect of the letter will be minimal if the reader does not understand the content because of technical wording or the explanations are more appropriate for other engineers.
- Keep it simple, stupid (KISS)! The length of the letter is important. A short letter has a greater chance of being published and then read, so follow the word limit guidelines of the paper.
- Include your professional affiliation if it is appropriate and permissible. This can add credibility to your information and opinion.

- Make your points relative to the interests of the readers. The editor will be more likely to select your submission if he or she believes that the readers will be informed or challenged by the content.
- If you are challenging the accuracy of a recent article in the newspaper, make sure that your facts are accurate and unbiased and the wording in your letter unambiguous.

9.4.2 INTERACT WITH POLITICAL REPRESENTATIVES

Letters to political representatives are an alternative to writing a letter to a newspaper editor. This is a good choice if the representative is currently in a position to act on a public policy issue that affects engineering or will be impacted by engineering knowledge, such as a bill to provide additional funds for infrastructure or to reduce harmful emissions that contribute to global warming. While letters to representatives can be longer than those to the editor, long letters should still be avoided. In these cases, it is permissible to have attachments to the letter, which is where details should be included. The letter will likely go through several levels of legislative aides before the legislator gets to see it; therefore, you will want to make compelling arguments that will make the aides believe that your letter is sufficiently important that the legislator should read the letter prior to voting on the matter. If the letter is in reference to a specific piece of pending legislation, include the number of the bill.

9.4.3 BLOGGING

Creating a blog is another avenue for participating in public policy debates. A blog can potentially reach a large audience, although some of the audience may be external to the region of interest. However, unlike letters to editors or public officials, blogs do not have a word limit, although if a reader decides not to read all of a lengthy blog, he or she may miss major points. Therefore, like a letter, the essay should be as brief as possible, but well organized and written at a level of the intended audience. A blog written for a civil engineering audience may be totally incomprehensible for a non-technical audience. A blog has the additional advantage that it offers the opportunity for feedback from readers.

9.4.4 INTERACTION WITH PUBLIC AGENCIES

Letters can also be written to officials of a relevant public agency. Generally, these officials will have greater knowledge of the technical details of the issue than a legislator. Thus, the letter may include a more detailed discussion of relevant technical ideas. However, the letter should still be accurate and concise. References to relevant articles in technical journals can add credibility to your arguments. It may be a good idea to have a colleague review the letter prior to mailing it. In the letter, it is appropriate to briefly summarize your background relative to the issue.

for American engineering students is absolutely essential to ensure long-term job security (Brown, 2005).

10.6   ACTIVE PARTICIPATION IN GLOBALIZATION

To actively participate in globalization, engineers must take several steps. First, engineers must consider the root of their knowledge: their engineering education. Having technical knowledge and competency is simply not enough to ensure job security. Engineers must find ways to thrive in an integrated, international environment. Engineering students should take advantage of extracurricular opportunities that will broaden their experience. This includes but is not limited to travel and study abroad, volunteer work with global organizations such as Engineers Without Borders, research opportunities, internships, and co-op positions. Taking courses beyond technical subject matter that emphasizes nontechnical aspects of professional engineering practice, such as leadership, communication, professionalism, teamwork, and public policy, will better prepare a student for working in a global job market. Competency in only the technical elements of the undergraduate civil engineering curriculum is an inadequate preparation for practice in a global professional environment. An equal level of competency in other components, like the humanities and social sciences, is part of the minimal education requirements.

Second, engineers must understand how globalization amplifies the effects of their work. Engineering decisions affect the economic and environmental health of both the engineer's own community and that at an international scale. The issue of the effects of transboundary air pollution was one example. The use and misuse of non-renewable natural resources affect both the country that uses the resources and other countries that must compete for the resources because they have a limited supply. Globalization must also consider long-term environmental effects and the results of both large-scale natural disasters and climate change. Today, civil engineers develop and implement international standards for both developed and developing countries, such as world health standards for adequate sanitation facilities and drinking water. Engineers from developed countries must be willing to transfer and adapt knowledge from their countries to developing countries that lack long-term experience in modern engineering.

Third, civil engineers must be actively involved in global professionalism. Civil engineers must meet cultural, language, legal, and political challenges and simultaneously make sure to respect cultural views and differences. Engineers also face the challenge of practicing ethically in a global marketplace. The engineer must be open to learning about other cultures' customs, mannerisms, and languages. Being educated in dealing with international partners and customers is crucial in today's global market.

Fourth, engineers must be ready for change. Experts in the civil engineering field agree that the demand for a new kind of civil engineer is growing, one that is flexible, broadly educated, and capable of meeting and tackling future challenges. The ability to use critical thinking skills in solving these new challenges is essential. Whether it is cultural, technological, social, or economical change, the civil engineer must be ready to adapt. An engineer who has considerable experience in his or her own country but

---

GLOBALIZATION: A CASE STUDY OF MATERIALS

Policy, legislation, and changes in technology have caused a need for more industrial wood sources in emerging economies. Many other materials are relevant to globalization because of the market demand. The steel industry in the United States is just one example. The distribution of raw materials is one factor, but the globalization of engineering materials will need to be addressed.

The International Institute for Applied Systems Analysis (IIASA) Forestry Program predicts that in 2030, African industrial pulp log production will increase from 20 million m³/yr to as much as 90 million m³/yr, and industrial saw logs will increase from 35 million m³/yr to as much as 150 million m³/yr. The rise in production will result in spite of the tight supply and increasing demand that is occurring on a global scale. China, in particular, is a major player in industrial log imports: China imports 2.5 million m³ of roundwood equivalents of forest products annually from Africa. Areas such as the European Union and India are also demanding more forest products and increasing the pressure and exploitation of Africa's forest resources.

In addition, the demands for wood and biomass used for energy production will increase. In the 2003 African Outlook Study, the Food and Agriculture Organization predicted that in 2030 wood production for fuelwood and charcoal will be 200 million m³, four times higher than the annual production of firewood. The total projected utilization of wood, both for industry and energy consumption, was estimated to be 1 billion m³/yr, compared to 650 million m³ today (Nilsson, 26–27). Forest product production is especially relevant to civil engineering because of the increasing importance of energy.

---

is willing to learn about others' cultures may find that the experience is inadequate. Different conditions may make self-study necessary in order to adapt the experience so it is helpful in the other country. All of this is part of the need to change.

10.7 DISCUSSION QUESTIONS

1. Compare and contrast the business customs of the United States and another country outside of North America. Then develop a set of global customs that would be acceptable in both countries.
2. What is job outsourcing? What are the advantages and disadvantages of outsourcing jobs? Give three examples of civil-engineering-related tasks that could be outsourced to another country.
3. Identify a major multinational civil engineering project other than Burj Dubai. Describe how globalization could play a role in the design and construction of the project. This may include procuring materials, engineering expertise, or knowledge from other areas and countries.

4. Identify a pollutant chemical source (e.g., DDT, mercury, PCBs, ozone) that has or has had a global impact. Identify the laws and regulations, if any, that have been put into place to prevent or reduce the pollution in two different countries. Were these laws and regulations successful? Why were the laws different?

5. Describe the negative impacts of globalization in developing countries. Consider poverty and access to clean water, sanitation, education, and medical care. How could civil engineers play a role in overcoming these negative impacts?

6. Briefly describe the North American Free Trade Agreement (NAFTA) and the General Agreement on Trade in Services (GATS). How do these relate to globalization and engineering?

7. Assume that in another country under-the-table payments are an accepted way of securing contracts. Identify elements of such a business practice that would violate the ASCE Code of Ethics.

8. What should a civil engineer who is practicing in another country do if he or she recognizes that nonrenewable natural resources are being wasted during construction activities?

9. Why would different countries have different criteria for licensure? What are the benefits of having common criteria?

10. Discuss the benefits and drawbacks of retaining the services of a civil engineer who is native to the country in which you are bidding for a project.

11. What criteria could be used to evaluate the potential benefits of a civil engineering company becoming involved in multinational work?

12. Select a culture and identify several differences between it and U.S. culture. Discuss how each difference might influence civil engineering practice and possible ways of overcoming the difference.

13. Identify three world health issues and discuss how each is relevant to civil engineering practice.

14. Select a nonrenewable natural resource and discuss the role that civil engineers could play in its preservation.

15. It has been reported that under-the-table payments are a common business practice in some cultures. Discuss why such payments are considered unethical in engineering practice in the United States and guidelines that should be provided to U.S. companies that practice in such cultures.

16. Discuss why natural resources should be viewed as global resources.

17. Identify ecological aspects of forests and why forest ecology is a global issue.

18. To what extent should a global organization like the United Nations be able to intervene in the environmental affairs of a member country?

19. Why is global environmental health an ethical issue for civil engineers?

20. If one country that is relatively rich in a particular nonrenewable natural resource does not practice sustainability with respect to that resource, how might this affect global sustainability?

10.8 GROUP ACTIVITIES

1. For one of the following issues, discuss the global aspects of the issue from an economic, cultural, professional, geographic, and environmental perspective.

   - *Air pollution*: Gaseous discharges emitted from manufacturers in one country travel throughout the globe.
   - *Water quality of transboundary river flows*: Pollution generated in one political jurisdiction and discharged into transboundary rivers can cause environmental damage or economic hardship in another jurisdiction.
   - *Different ethical standards*: Engineers who practice in different cultures or political systems may have different ethical standards for practice. How can these be globalized?
   - *Differences in competency*: Engineers may be asked to perform engineering services in localities where conditions differ (e.g., an engineer licensed in Colorado providing services in a region subjected to tsunamis). How does local licensure impede globalization?

2. As engineers become more globalized, why must they be more aware of engineering ethics? Describe the advantages and disadvantages of a global code of ethics.

REFERENCES

Agrawal, A., Chatre, A., and Hardin, R. 2008. Changing governance of the world's forests. *Science* 320(5882):1460–1462.

Barboza, D., and Bradsher, K. 2006. Pollutoin from Chinese coal casts a global shadow. *The New York Times*. Retrieved from http://www.nytimes.com/2006/06/11/business/worldbusiness/11chinacoal.html

Brown, J. L. 2005. Wanted: Civil engineers. *Civil Engineering*. 87(6):46–49.

Friedman, T. L. 2006. The world is flat. New York: Farrar, Straus & Giroux.

Krane, J. 2006. Workers riot at site of Dubai skyscraper. [Electronic version]. *Associated Press*, Retrieved from http://www.breitbart.com/article.php?id=D8GGPJJG3&show_article=1

Reina, P., and Post, N. 2006. Final height of "on-deck" tallest tower shrouded in secrecy. [Electronic version]. *McGraw-Hill Construction*. Retrieved from http://www.construction.com/NewsCenter/TechnologyCenter/Headlines/archive/2006/ENR_1106b.asp

The tower—facts and figures. 2009. Burj Khalifa. Retrieved from http://www.burjkhalifa.ae/the-tower/fact-figures.aspx

U.S. Department of Labor, Bureau of Labor Statistics. 2007. *Occupational outlook handbook: Engineers*, Retrieved from http://www.bls.gov/oco/ocos027.htm

11 Leadership

CHAPTER OBJECTIVES
- Identify principles and attitudes of leadership
- Provide guidelines for leadership evaluation
- Discuss values and ethical issues relevant to leadership
- Identify ways of gaining leadership experience

11.1 INTRODUCTION

You are part of a group assigned a project in one of your classes. If you were assigned to be the group leader, what do you imagine the group members would expect of you? They would likely expect you to plan the required tasks, set up a timeline (i.e., a schedule of dates on which tasks need to be completed), make assignments, follow up to ensure each person in the group completes his or her tasks on time, and organize the group report. What would happen if you failed to complete these tasks? A less than acceptable report is quite likely because of ineffective leadership. The quality of the final group report will depend on the ability of the group leader to organize, plan, and manage the individual tasks.

The field of civil engineering requires strong leaders, as society trusts the products and services that engineers provide. People assume that civil engineers have properly designed the facilities that they need. In the past, strong leadership has provided the public with safe drinking water from the sink and the use of safe roads and bridges to get to work. However, when a bridge fails, such as the failure in Minneapolis in 2007, the public may question the safety of other public facilities. It is the responsibility of civil engineers to maintain the trust that the people place in them. Meeting this responsibility requires strong leadership. Faced with our global environmental and climate crises, civil engineers will need to provide leadership in meeting goals of sustainable development and a clean environment.

11.2 LEADERSHIP: A DEFINITION

Leadership is often thought of as the ability to direct the behavior of others, as a boss would assign tasks to his or her employees. However, leadership is much more than directing the activities of employees. The responsibilities of a leader go well beyond the internal workings of the organization. Leaders have multifaceted responsibilities, including setting organizational goals, interacting professionally with the community, and acting to influence the direction of the profession.

Often engineers are practical, analytical, intelligent, and nonemotional; they are viewed as lacking a humanistic perspective and would rather deal with tangible things instead of people (Florman, 1976). Engineers must broaden their skills and qualities to include emotional and humanistic aspects. Leaders, especially, should be aware of the needs of the organization and must view the end product as being more than just a tangible product, but also an engineered product that will fill the needs of the community. For example, a company designing a center for autistic children must be aware of the needs of the children and their families, by working with various autism resources and groups to create a viable design. As illustrated by this example, the end product is not just the building; it is a place where autistic children and their parents can receive treatment and support and reach their potential as citizens.

Leadership in civil engineering can then be defined as the ability to create a vision, direct an organization or team, motivate subordinates, and efficiently use the resources required to plan and execute an engineered design that fulfills both the physical and emotional needs of the community. Does every civil engineer reach the leadership plateau? No! Unfortunately, many who have the potential fail to achieve a position of leadership because they take the narrow view that success depends solely on technical accomplishments. Sometimes this results from a lack of mentoring. In other cases, a lack of confidence prevents the person from seeking responsibilities that prepare him or her for leadership.

The leader must constantly set an example that he or she is committed to positive innovation, and that he or she is capable of overcoming obstacles and embracing change when needed. Moreover, this attitude should be spread throughout the entire management team and to the team members (Adair, 2005).

11.3 PRINCIPLES OF LEADERSHIP

A leader's success depends on his or her adherence to principles of leadership, which include (1) accepting responsibility for planning and organizing resources, including human, financial, and physical; (2) having the ability to inspire innovation and creativity in others; (3) having an interest in motivating others within the organization; (4) demonstrating the ability to provide the organizational structure needed to achieve organizational goals, especially in the face of changing conditions; and (5) accepting the responsibility for making organizational decisions. Planning encompasses both short-term and long-term objectives, establishing policies, forecasting future demands, and understanding potential uncertainties. To remain on the forefront of any industry requires continual change through innovation and creative activity. The leader must find ways to foster innovation, even encouraging personnel to independently seek innovative solutions to problems that they recognize. Ideally, each person in the organization is self-motivated; however, a leader can motivate employees by matching the assignments to the interests of the employees. The organizational structure is a plan on the way that organizational resources are best used. The organization chart identifies who is assigned decision authority and responsibilities. The leader must identify alternative solutions to organizational problems and ultimately make the decision.

CASE STUDY

The American effort in building the Panama Canal was led by Lt. Col. George Washington Geothals, U.S. Army Corps of Engineers. He practiced all of the principles of leadership. He planned and organized the activities and had his command ensure that the activities were carried out. His innovative outlook was responsible for building locks large enough for the boats of the time. Geothals' vision to use intermediate gates within a lock led to greater efficiency. Geothals had a Sunday morning, open-door policy during which workers could meet with him; this increased worker motivation and job satisfaction. In summary, Geothals was very successful because he had the experience necessary to learn and apply the principles of leadership (Parker, 1986).

11.4 ATTITUDES AND SKILLS OF LEADERS

Sports team managers, church elders, university deans, city mayors, student chapter presidents, and the alpha-males of wolf packs are usually viewed as leaders. What characteristics do they have in common if they are truly leaders? According to the ASCE *Body of Knowledge*, leaders need to have the following skills, abilities, and characteristics:

- *Adaptability*: The ability to change in response to new demands or constraints.
- *Agility*: Skillful under pressing conditions.
- *Commitment*: A sense of duty to fulfill responsibilities.
- *Competence*: Having the knowledge, skills, and abilities to achieve an end.
- *Confidence*: A feeling of assurance about the ability to succeed.
- *Courage*: Steadfast in meeting adversity.
- *Curiosity*: The internal motivation to learn new things.
- *Discipline*: The ability to control behavior to meet specific goals.
- *Enthusiasm*: An eagerness and the ability to inspire the interest of the group.
- *Industriousness*: Diligently active in pursuit of organizational goals.
- *Initiative*: The instinct to initiate action, including motivating subordinates.
- *Integrity*: Rigid adherence to a code of behavior that serves group goals.
- *Persistence*: An attitude to persevere in the face of obstacles.
- *Selflessness*: Having greater concern for achieving group goals than for meeting personal goals.
- *Vision*: Intelligent foresight related to group matters and being perceptive about future trends.

Being successful as a leader does not require outstanding performance or ability in all fifteen of these skills, abilities, and characteristics. However, self-assessments should identify a person's weaknesses in this list. Self-improvement through lifelong learning should be made when someone detects a weakness in any of these skills, abilities, or characteristics.

11.5 VALUES IN LEADERSHIP

Employers seek graduates who possess the leadership skills that will allow them to utilize their technical knowledge. Success as an engineering leader depends less on technical ability and more on the coordination of these skills. Leaders must also have a value system that will allow them to lead efficiently and effectively. Values that are important to a leader include:

- *Integrity*: Adherence to technical and moral codes of behavior.
- *Respect*: Consideration for the well-being of others and willingness to listen to ideas.
- *Assertiveness*: Enforcing rules with the proper authority.
- *Honesty*: Being trustworthy and capable of acting without deception or fraud.
- *Creativity*: Willingness to think outside of conventional routes.
- *Dependability*: Being responsible and reliable, able to complete tasks on time.

In addition to fulfilling the needs of the organization, the leader also has responsibilities to the local and global communities. An ideal leader should possess excellent planning and organizational skills, be trustworthy, and be self-motivated. Leaders should be open to the input and needs of other people and should motivate subordinates toward a common goal.

The group must feel respect, a likeness, and loyalty toward the leader. They must be trusting of his or her ability to lead the group. The leader should provide subordinates with the opportunity to ask questions and voice concerns. Communication between the leader and the group is vital to a project's success. The leader should keep an open line of communication with the group members by having an open-door policy.

CASE STUDY

After enlisting and serving as an engineering officer in the Civil War, Washington A. Roebling returned to help his father construct the Cincinnati and Covington Suspension Bridge, the longest suspension bridge in the world at that time. Later, John Roebling, Washington's father, accepted a job as chief engineer on the proposed Brooklyn Bridge. John died in a freak accident during the early stages of construction of the Brooklyn Bridge. Washington was left with his father's company and the responsibility of building what was then the largest bridge in the world. Washington's wife, Emily, worked alongside him as an assistant. Emily's social skills and intelligence helped overcome many of the problems that the company faced, including problems with materials, politics, and public relations.

The most difficult construction obstacle of the Brooklyn Bridge was the pneumatic caisson foundations. Washington Roebling gained competency in caissons through self-study. Washington spent more time in the caissons than any of his assistants, thus serving as a role model. He contracted decompression sickness, also known as caisson disease, which he developed because of the pressure differences inside of the caissons. Washington was left paralyzed, partly blind and deaf, and unable to speak normally. After the near-death experience, Washington spent the winter away from the site and prepared detailed directions for the construction of the cables and the structure. When Washington's ability to head the project was questioned because of his deteriorating health, Emily spoke to several groups, including ASCE, and successfully won over officials with her tact, diplomacy, and communication skills. Emily studied and mastered civil engineering topics to assist her husband.

Washington continued to direct the construction of the bridge from his sickbed. Washington often watched the construction through his room window with binoculars, while Emily became his main assistant. Emily inspected the construction site daily and relayed Washington's messages and information to the workers.

Throughout most of the construction, Emily was the Roebling face that the public saw. Emily took the initiative to gain the engineering skills necessary to build the most legendary bridge of its time. Her efforts in overseeing the construction were recognized at the opening ceremony of the bridge, when New York's congressman awarded both Washington and Emily the title of *chief engineer.*

Both Washington and Emily Roebling were excellent, dependable leaders. They both went beyond their call of duty. Both Roeblings had immense integrity and adhered to both the technical and moral codes of behavior. Washington still led the construction of the bridge, even from his sickbed, which showed high levels of courage, persistence, and commitment. Emily, who was, without a formal education, self-taught and had mastered civil engineering topics, stayed by her husband's side throughout the project. This showed adaptability and discipline. Both felt morally obligated to complete the bridge for the sake of society's needs. Their selfless attitude and commitment to the project allowed them to lead and excel (Weingardt, 2004).

11.6   ETHICAL ISSUES OF LEADERSHIP

A leader has many and varied responsibilities: hiring and firing, setting organization goals, representing the organization in community outreach, providing financial control, including generating business, and not least, ensuring guidance on ethical matters. A leader's responsibilities related to ethics and professionalism include:

- Identify ethical and social responsibilities for all employees.
- Demonstrate the conduct expected of employees through one's own actions, i.e., serve as a role model.
- Use organizational resources to promote ethics education.
- Articulate the consequences of violations of ethical standards.
- Use the power afforded by the position to enforce proper conduct.

Failure to fulfill these responsibilities can cause the organization to fail.

As indicated, one responsibility of a leader is to serve as a role model, which requires the leader to be seen by subordinates and peers as being honest. This involves having a clear understanding of professional ethics, the ability to communicate, the ability to recognize ethical issues, and the ability to resolve ethical dilemmas. An effective leader must also possess the desire, willingness, and courage to resolve ethical conflicts within the organization. Conflicts that arise because of differences in values, which is often referred to as office politics, can be quite contentious and contribute to inefficiency of organizational operations.

Leadership does not always fall to the person that the organizational chart indicates is the top person in the company. Leadership is needed at all levels of an organization. Engineers at all levels often need to provide leadership when professional or ethical conflicts occur even at the project team level. Professional issues may arise when working with fellow engineers or interacting with management. Depending on the size and type of work environment (e.g., a new start-up firm, a nonprofit organization, a large international corporation, or a military organization), management and professional expectations may differ, but leadership on value, professional, and ethical issues will always be required.

The failure of one team member to provide a fair share of work is a common example of a professional issue that requires leadership. If one team member is viewed as failing to contribute, then the team leader must address the issue in a timely manner; otherwise, the efficiency of the team will suffer. If the leader does not recognize the situation or fails to act on the lack of adequate effort by one member, then it may be another team member's responsibility to address the issue directly with the member or report the situation to the leader. In many cases, a team member must take the initiative to handle the problem. This initiative is often difficult to take because the engineer may feel like it is not his or her place to tackle the problem. A person with strong leadership skills will not ignore the responsibility to act.

One of the ethical issues that engineers must address is protecting public health, the quality of life, and the environment of their community. While ecoleadership was not a role to play even a decade ago, the importance of sustainability and the environment has added a new dimension to leadership in engineering. This responsibility extends from the local community to the global community. For example, soil eroded from a construction site in Ohio can end up in the delta of the Mississippi River, which illustrates the effect that engineering activities can have hundreds of miles from the site. Similarly, dump trucks can emit chemicals that contribute to air pollution locally or regionally. The engineer in management must provide ecoleadership to minimize such problems. A leader must also balance his or her personal integrity and standards with those of the management or company, as these may possibly conflict. One example of an ethical issue that a leader may have to confront is in the final stages of product design. Assume that a product design has been approved and is ready to go into the manufacturing phase. The leader notices that there is a flaw in the design, one that was overlooked by all of the design team and by the staff who approved the design. The flaw is small, though it could cause a possible safety recall in the future. Management may want to begin marketing the product, with the argument that the flaw is minor, while any delay would result in a significant economic loss. The engineer may feel an ethical responsibility to fix the design flaw prior to its distribution. Obviously, the values of the engineer and management differ. The engineer may believe that alerting management to the problem is his or her only responsibility. At the other end of the spectrum, the engineer may feel the issue sufficiently important that he or she will quit if management decides against correcting the design flaw. Cases where individual and management values are in conflict are not rare.

In some companies, codes of conduct are developed by management. These are unfortunately developed after an ethical problem arises. A code may address issues such as not using company resources for personal gain, adhering to confidentiality rules, and not falsifying information. When a leader is faced with an uncertain value conflict among subordinates, the written code of ethics can provide both guidance and support for any decision. A leader should understand that codes are of value and initiate the development of a code before an ethical conflict arises. This will help the leader gain credibility and trust from the group members.

11.7 GAINING LEADERSHIP EXPERIENCE

Leadership is both a learned and an applied skill. A civil engineering student can gain leadership skills at the undergraduate education level. A student can take a leadership role in design teams, competitions, clubs, and societies on campus. Excellent examples of these opportunities include being a leader in an introductory engineering project or running for an officer position in a student chapter.

The student must keep in mind that a leadership role, especially in an established group, may not be available immediately. Many times, a student must prove his or her loyalty, interest, and ability to be a leader by actively participating in the group over an extended period of time. In Engineers Without Borders, where groups of students plan and execute engineering-related international work, all students who are involved in a project do not go on the trip: it is simply not feasible to fund airfare for a group of fifty students to do the work that ten students can do, where the money could be applied elsewhere. To get a spot on the trip, the student must show that he or she has the technical capability and is committed to the group and the project. A student does this by attending meetings regularly, contributing to the design, working with other students outside of meetings, and participating in fundraising at sports events, among other activities. Similar to this, an officer position in a society or club may not be immediately obtainable. Students should take advantage of short-term leadership opportunities that will showcase their ability and commitment to the group. This work might include helping to organize a fundraiser, filing paperwork, creating advertisements, and recruiting new members. Participating in such activities allows others in the group to recognize the student's commitment and attitude.

The tasks in which students participate as they rise to a position of leadership are largely not technical in nature but provide opportunities to use the same leadership skills as corporate CEOs. The tasks generally involve administrative duties and directing the activities of other group members. Motivating subordinates is a primary responsibility of a leader, and a student chapter officer learns about motivating fellow students to show up on time at fundraising activities. Leaders are generally good communicators, and a student chapter member can improve his or her ability to communicate by directing committee meetings. Leaders of engineering firms must plan organizational activities, and a student who organizes an effort at concrete canoe building will gain planning experience that will be of value as he or she rises to a position of leadership in an engineering company.

Success in a leadership position is often the result of good mentoring in the early part of one's career, even while an undergraduate. Every undergraduate should have at least one mentor. An upperclassman can be helpful in knowing which faculty members act as good mentors. A faculty mentor can identify activities and provide opportunities that offer long-term advantages, such as getting involved in research, being active in student organizations, selecting internships, and taking nontechnical electives that will have the benefits needed to become a leader. Even while on a summer job, an undergraduate should discuss with engineers civil engineering as a career path and the activities that they found to be most valuable in their careers. The undergraduate student can gain knowledge of leadership from the experiences of past leaders.

11.8 LEADERSHIP IN AN ENGINEERING CAREER

Developing leadership attitudes and skills should begin early in one's career. Even during the first year or two after one's undergraduate matriculation, observing leaders at the place of employment can provide valuable guidance. The leadership performance of senior staff members can provide guidance on the proper conduct of a leader. Where senior staff do not function well, their negative performances can serve as valuable guidance on conduct to avoid when serving in a leadership position. This is a time in one's career when knowing how to evaluate leadership can be a worthwhile education (see Section 11.10).

While observing leadership performance can be instructive, it cannot replace actual experience. Leadership experience can be obtained both at the place of employment and externally in other professional activities. A young engineer should seek advice from his or her mentor within the company about assignments that will provide leadership experience. Outside the company, professional group activities at the local or national level offer the opportunity to gain leadership experience. This generally requires one to volunteer his or her time on behalf of the group. Companies often support such activities, as it is viewed as good exposure for the company. While these leadership activities may not occur at the start of one's career, active participation prior to taking on a leadership role is often necessary.

Leadership development must continue throughout one's engineering career. Companies will often provide support for professional development conferences, workshops, and seminars. Active participation in professional organizations at the local, state, or national level will provide leadership experience. Globalization and the increasing interconnectedness of communities around the world introduce new obstacles and opportunities for engineers to provide leadership. Active participation in professional committees that deal specifically with international participation and issues provides the opportunity to develop global contacts.

CASE STUDY

When Soichoro Honda, the founder of the Honda Motor Company, was a child, he saw an automobile coming down a country road and immediately raced after it in excitement. Years later, as an aspiring engineer, Soichoro sought to be an innovator during Japan's postwar period, when gasoline was rare. He created a motorized bike that used the oil of pine tree roots as fuel and a hot water bottle as a fuel tank.

Soon after creating Honda Motor Company, which produced motorized bikes, Soichoro was on the verge of bankruptcy because the company lacked the proper financial background and experience. Soichoro responded quickly by joining forces with businessman Takeo Fujisawa, and Honda's motorcycles were soon leading on racing circuits around the world.

The Honda Motor Company moved into the car production business for the mass market, using Formula 1 racing as the testing ground. When environmental problems caused by gasoline emissions became an issue, Soichoro Honda immediately recognized the opportunity to gain a competitive advantage. He used the talents of his company's young engineers to develop a low-emission, water-cooled engine. This engine was placed into the Honda Civic, which soon became a fuel-efficient, worldwide success.

The innovative and creative mind of Soichoro Honda created one of the leading automobile manufacturers in the world. He possessed many hallmarks of a great leader. He was flexible and willing to take risks, from tackling financial troubles to addressing a new environmentally concerned consumer market. Honda was aware of the ever-changing state of the world and employed his company with the task of seeing every obstacle as an opportunity instead of a problem, and using every opportunity as a chance to improve the lives of others (Adair, 2005, pp. 53–61).

11.9 CONSIDERATIONS FOR CREATING A NEW ORGANIZATION

Creating a new organization requires experience and knowledge. While books devoted entirely to the subject have been written, only a few elements of the task will be presented. Two aspects are especially important, planning and organization, which are defined as:

Planning: The process of identifying, analyzing, evaluating, and deciding among alternative opportunities for the purpose of meeting organizational goals.

Organization: The process of using resources (human and material) to execute the organizational plan.

Leadership is critical to these two responsibilities.

The purposes of planning include (1) minimize risk and uncertainty, (2) increase efficiency, (3) coordinate efforts within the organization, and (4) evaluate long-term consequences of decisions. While a number of representations of the planning process have been developed, the following is one model of the process: (1) organize goals, (2) develop the alternative actions to reach the objectives, (3) develop a decision process, and (4) develop plans to execute the selected alternative. The leader plays a critical role in the planning process, as he or she must direct the development of the organizational goals. To achieve this, the leader will need to have experience in planning, knowledge of the organization, good communication skills, and an understanding of the factors that affect the growth of the specific industry. The planning must be flexible, and change is necessary when warranted by economic or technological change.

Organization also requires strong leadership. The activities in the organization phase include (1) develop a formal management structure and select a leader for each department, (2) motivate employees, (3) define job responsibilities, and (4) coordinate departments.

11.10 EVALUATION OF LEADERSHIP

Books on management include numerous theories that have been proposed to evaluate a leader's effectiveness. Unfortunately, none of these are universally accepted. General evaluation criteria can be listed, but all of them are not important in every case. Therefore, the criteria important in a specific case must be identified, ranked in importance, and then applied to the valuation of a specific leader. The following evaluation criteria are common to many of the theories on leader effectiveness:

- *Technical competency*: In order to delegate tasks, the leader must appreciate the abilities that will be needed to complete the assignment. Technical competency is therefore necessary to assign tasks and supervise projects.
- *Ethical maturity*: Identifying values important to an organization and resolving ethical conflicts requires ethical maturity; a person is ethically mature (see Section 15.9) if he or she understands the breadth of his or her ethical responsibilities and makes decisions where the ethical responsibilities to completing entities are properly balanced (i.e., unbiased).
- *Decisiveness*: Decision making is an important responsibility of a leader. An effective leader will be one who is decisive and in hindsight continually made the correct decisions.
- *Motivation*: To achieve organizational goals, a leader will need to ensure that subordinates are motivated. This requires supervision, good communication skills, and the belief that the organization is sensitive to the workers' needs.

11.11 DISCUSSION QUESTIONS

1. How does working or volunteering improve one's leadership skills?
2. Describe an experience that you had during an engineering design project. Was there a leader? Was the leader self-appointed or elected? How did the leader interact with the group and was he or she effective?

CASE STUDY: EVALUATION OF STUDENT CHAPTER OFFICERS

Each civil engineering department has at least one student chapter that elects officers. During any one semester, the success of the chapter depends heavily on the quality of leadership provided by the student chapter president and other officers. What criteria should be used to evaluate their performance? The following is a brief structure for an evaluation of the officers, with the general criteria given and the specifics then identified:

- *Plans well*: Stays within budget.
- *Good organizational structure*: Committees were developed in a timely manner (at the start of their administration).
- *Good supervisory skills*: Meets regularly with committee chairs.
- *Motivates members*: Committees complete tasks on time and meeting attendance is good.
- *Acts ethically*: Does not use chapter funds for personal gain or even for questionable activities, such as pizza during officers' meetings.
- *Serves as a role model*: Grooms younger members to take leadership in the future.
- *Communicates well*: Holds regular meetings.
- *Develops external relationships*: Invites speakers from CE firms to meetings.

While other criteria could be developed, the specific criteria selected will depend on the purpose of the group. The importance of each criterion should be identified before actually conducting the evaluation.

---

3. What is the difference between management and leadership in civil engineering?
4. How might a leader make himself or herself seem more open and responsive to group concerns? What methods could the leader utilize? What are events that the leader could create?
5. Earlier in the chapter, fifteen skills, abilities, and characteristics were defined. Assessing yourself, identify the five that you feel most confident that you have as a leader. Also, identify the five that you believe are your weakest. Briefly discuss how you could improve with these five weakest.
6. Apply the definition of leadership to a capstone course group. Identify and discuss three principles of leadership that would be especially relevant to the capstone team leader.
7. From the perspective of a local community, which principles of leadership are important? Select two principles and discuss why they are relevant to a community.
8. Identify one of the principles of leadership and discuss how serving as a student chapter president while enrolled as an undergraduate can serve to develop this leader's potential.

9. Assume that you are project leader for an engineering materials laboratory project. Identify one principle of leadership that is especially important to project success and discuss reasons for its importance.
10. Identify and briefly discuss the tasks involved in organizing a group project.
11. What criteria could be used to evaluate the leadership of an organization? Place the criteria identified into three groups: very important, important, and least important.
12. Identify ways that being a leader of a group of civil engineers would differ from leading a group that included an accountant, two sociologists, a lawyer, and two civil engineers.
13. Discuss why serving as a role model is an important part of being a leader.
14. Why is mentoring important in the role as a leader?
15. What are the responsibilities of a mentor?
16. A civil engineer is leading a group that is investigating failure of civil engineering systems during a natural disaster. Which principles of leadership would be especially important? Provide relevant reasons for your choices.
17. Why are leaders important to organization or group success?
18. How does someone gain the experience necessary to be a leader?
19. Contrast autocratic and democratic leadership styles.
20. What is meant by leadership flexibility?
21. How can a leader motivate an employee who is just interested in receiving a paycheck?

11.12   GROUP ACTIVITIES

1. Expand upon the case study on Washington and Emily Roebling by examining the actions that demonstrated good leadership.
2. Use the fifteen attitudes and skills listed in this chapter to develop an evaluation form for use in evaluating the leadership of a president of a student chapter.
3. Create a list of leadership characteristics and a second list of personal characteristics. Discuss the extent to which the elements of the lists agree.
4. Develop a plan for creating a new student chapter that focuses on global sustainability. Outline its organizational goals, a timeline of plans for getting the group started, and a framework for leadership.

REFERENCES

Adair, J. 2005. *Leadership for innovation: How to organize team creativity and harvest ideas.* Philadelphia: Kopan Page Limited.
Florman, S. C. 1976. *The exitential pleasures of engineering.* New York: St. Martin's Press.
Parker, D. S. 1986. Leadership on historic mega projects. Presentation at the Department of Civil Engineering, University of Maryland, College Park, April 22.
Weingardt, R. G. 2004. Clifford Milburn Holland and Emily (Warren) Roebling. *Leadership and Management in Engineering* 4(3):116–119, July 2004.

12 Teamwork

> **CHAPTER OBJECTIVES**
> - Identify values and factors important to team success
> - Identify activities for gaining experience in teamwork
> - Discuss the roles of time management, communication, motivation, and creativity in teamwork
> - Provide criteria for evaluating team performance

12.1 INTRODUCTION

The word *team* usually brings sports to mind—the Yankees, the Packers, the 76ers. Maybe *team* brings to mind a group of draft animals yoked together for farmwork. If the centerfielder does not chase down fly balls, then the other eight players may not be able to compensate and the team will lose. If one of the draft animals is pulling in a different direction, then the wagon will not get pulled in the right direction. The same is true for a business team. Success depends on the extent to which the group is organized and works together. One slacker on a business team can lead to failure for the entire team.

Teams are utilized in almost every aspect of a civil engineer's career. On large projects, design work involves structural, geotechnical, and environmental engineers. This group reflects an intradisciplinary group. In a broader context, the final design will depend on interactions with surveyors, planners, lawyers, and accountants. This reflects a multidisciplinary design team.

Civil engineers must be able to function effectively on a team. This requires an understanding of the values relevant to teamwork, a diversity of knowledge and experience, recognition of the importance of time management and communication, a commitment to change, and the abilities of innovation, motivation, and creativity. These factors are important to team success and serve as criteria for team effectiveness.

As stressed in the ASCE *Body of Knowledge*, technical competence of team members is important, but experience in teamwork is necessary for success. Civil engineering students can gain this experience at the undergraduate level. Many opportunities are offered in the classroom, but students will ultimately have to seek extracurricular opportunities to gain both intradisciplinary and multidisciplinary team experience.

12.2   VALUES RELEVANT TO TEAMWORK

Team effectiveness depends on the technical ability of the individuals, but more so on the degree of cooperation that is developed among team members. They need to be honest with one another, able to rely on and trust one another, and accountable. Therefore, values are central to team effectiveness. The following values are especially relevant to the success of teams:

- *Honesty*: Condition of being trustworthy and ethical, acting without deception or fraud.
- *Reliability*: Dependable for obligations and duties.
- *Efficiency*: Producing effectively with a minimum amount of wasted resources.
- *Accountability*: Taking responsibility for one's conduct; answerable for obligations and duties.
- *Promptness*: Being punctual and able to consistently meet deadlines.
- *Trust*: Reliable, integrity, ability, and character.
- *Diligent*: Industrious; perseveres in completing assignments.
- *Tolerant*: Able to stay motivated during conflicts with others.

Each value is vital in relation to the individual on the team and the team as a whole. The individual must display these values on a regular basis: he or she must be honest, reliable, efficient, willing to be held accountable, prompt, and trustworthy. Selflessness rather than selfishness must be the norm.

The engineer needs to prove himself or herself capable of doing the job effectively and on time, in order to gain the trust and respect of leaders and fellow engineers; however, for the team to succeed, these values must be embraced by each member. Even just one team member failing to adopt these values can lead to failure of the team effort. Accepting these values will enable the team as a whole to be efficient and effective, and will allow the team to complete the project while meeting time, money, technical, and ethical specifications.

Students should also be able to accept and integrate diversity, including differing perspectives, knowledge, and experience. With the spread of globalization and projects that involve people from different countries, team members must be aware that other members may have different cultures and expectations. In addition, teams that involve civil engineers work with others with different knowledge bases, including those from other engineering disciplines, the business world, communication specialists, and even politicians. Civil engineers may work directly or indirectly with people of different levels of education and expertise, ranging from field workers with little or no college education to CEOs of corporations. Civil engineers may also work with both senior-level engineers and recent engineering graduates on projects.

12.3   GAINING EXPERIENCE AT THE UNDERGRADUATE LEVEL

Civil engineers often work on two different types of teams and should be able to function within both. The first type is an intradisciplinary team, which consists of members in the civil engineering discipline, for example, structural engineers working with geotechnical engineers. Intradisciplinary work is stressed at the undergraduate level, where students are involved in design projects, lab exercises within a course, and capstone design courses. Other opportunities include joining civil engineering honor societies or an ASCE student chapter. A team whose objective is to build a concrete canoe is an intradisciplinary team even though a variety of experiences are necessary to be successful.

Multidisciplinary or cross-disciplinary teams consist of members in different engineering fields or professions, for example, civil engineers working with electrical engineers or community engineers working with public policy officials to plan a community development. Multidisciplinary work is not typically stressed at the undergraduate level, though this experience is vital to a civil engineering profession. Civil engineering students can gain experience by electing to take non-civil-engineering courses for upper-level technical electives (e.g., business, humanities). Extracurricular opportunities should also be pursued, such as a student government association, civil and service organizations, engineering societies such as Tau Beta Pi and Society of Women Engineers, and internships. Projects like Solar House or Solar Car Race are examples of multidisciplinary teamwork. It is especially worthwhile to be active on a team that goes well beyond engineering disciplines, such as those who work on projects for the community, e.g., Habitat for Humanity. While participating in any group activity, it is important to observe positive factors that contribute to successful operation and negative factors that keep a team from efficiently meting its goals.

12.4   TEAM FORMATION AND EVOLUTION

For purposes herein, a *team* can be defined as a group of two or more people working together to solve a problem in which professional engineering issues are involved. The group may consist entirely of civil engineers, which is referred to as an intradisciplinary team, or engineers and stakeholders outside of the profession, including political representatives, homeowners, or members of special interest groups; the latter group is referred to as a multidisciplinary team.

An important initial step in organizational decision making involves team building, which can be loosely defined as initial activities that encourage developing group cohesiveness. Team building will increase team efficiency, encourages continued communication, and should make the task a more positive experience for the team members. Tasks central to team building include

- Develop a full understanding of the problem
- Agree on the general goal and specific objectives
- Identify a leader, unless organizational management selected the leader before the team met
- Ensure that the needed expertise is available with the team composition
- Form subgroups to work on specific objectives
- Establish a timeline
- Identify lines of communication between the team leader and subgroup leaders

The value of team building should not be underestimated, as it is often critical to success.

The leader of the team may be selected by the organizational leadership, such as the CEO or business manager of a company, or someone from outside the profession, such as a political group. In some cases, the team might meet and agree on a member who will take on the leadership responsibilities. The leader has the responsibility to direct the interactions, activities, and feelings of team members. The leader may decide how allocated resources will be used. In the early stages of the team activity, the leader should try to develop team unity, i.e., a feeling of cohesion around the goal of the team. The leader will need to decide on his or her leadership style: democratic or autocratic. This will depend on the composition and experiences of the individual members.

A team goes through various phases en route to achieving its goal. First, the leader must provide an orientation, including establishing important interactions and ensuring that everyone knows their responsibilities and the general timeline for completing their work. Second, the team will develop a list of alternative plans for solving the problem. Third, the team will evaluate each alternative listed and select the one that seems likely to be the most effective. Fourth, the group will assess whether the resource allocation and timeline are appropriate. If not, they should recommend changes, which the team leader will need to direct. Fifth, the team should establish an evaluation plan. This will be formative in nature. It is especially important to establish the specific evaluation criteria so that each team member will know the metrics that will be used for his or her assessment.

12.5 FACTORS IMPORTANT TO TEAM SUCCESS

The success of a team is not a given. Some teams succeed, while others fail. Even within an organization, the degree of team success is variable. The following lists some factors that influence team success:

- The team goal must be clearly stated and understood.
- The team size is important, with the size depending on having the needed technical expertise and a reasonable workload per member.
- The chairperson must know how to organize the team, be unbiased with respect to the project goals, and be an effective communicator.
- Meetings must be well prepared and run, with all necessary follow-up work completed in a timely manner. Agendas are critical to meeting efficiency.

Success may be judged by outsiders solely on the extent to which the team goals were met. Team members themselves may have a broader view of success, as they will include factors such as the level of team harmony throughout the process and the extent that they felt that their time was well spent.

Characteristics important to the success of teamwork include leadership, values, team dynamics, and having the knowledge and experience necessary to meet all objectives. Team leadership has many of the same objectives as organizational leadership (see Chapter 11); however, the goal of teamwork is more narrowly focused to solving a specific problem. Values important to the success of teamwork were discussed in

Section 12.2. Communication (see Section 12.7), motivation (see Section 12.9), and creativity (see Section 12.10) are important elements of team dynamics.

12.6 TIME MANAGEMENT IN TEAMWORK

Time management is particularly important for civil engineers because each engineer plays a role in the completion of the final project. In addition, the progress of one engineer may depend on the progress of another. For example, an environmental engineer who failed to develop the water and soil database for a site in time may hinder the progress of the construction engineer.

Learning how to manage time effectively early on in one's undergraduate career will be extremely useful in the work field. College students especially may find time management to be a crucial skill to be able to succeed in the classroom and still have enough time for extracurricular activities. Time management is an excellent skill that can be developed through methods and practice at the undergraduate level. Some suggestions on how a civil engineering student can improve his or her time management skills include

- Actively using a calendar or planner to track dates of quizzes, exams, and deadlines
- When assigned a large project, breaking it down into manageable tasks and milestones
- Creating task lists (daily, weekly, monthly, yearly) and assigning priority to certain tasks
- For team meetings, determining the allotted time that the meeting will run beforehand

12.6.1 TIMELINE

Once the goal and objectives been defined, a timeline should be developed. A timeline is a schematic that identifies when various phases of the research will be met, as well as when deliverables are due and specific tasks are expected to be finished. A timeline is usually included with a research proposal and should be updated periodically as expectations are met or not met at the time specified on the timeline.

12.6.2 TO-DO LISTS

Experience has shown that to-do lists improve time efficiency. Productive individuals generally keep both a short-term (e.g., one day) and a long-term (e.g., weekly or monthly) to-do list. In addition to the specific tasks that need to be completed, efficiency can be improved if the priority of items on a to-do list is shown. The priorities help the individual focus on the more important tasks.

12.6.3 DAILY SCHEDULE

Maintaining a daily schedule is also a way of improving efficiency. Time slots of fifteen, thirty, or sixty minutes can be used. Some activities like classes, athletic practices, or family meals may not be under the individual's control. However, a considerable amount of a day may be under a person's control, and keeping a schedule will help him or her make the best use of the controllable time. Brief open periods should be devoted to doing simple tasks, with longer time slots saved for tasks that will require more time to complete. Both a schedule and a to-do list will help you measure your progress toward meeting your goals.

12.7 COMMUNICATION IN TEAMWORK

Communication is critical to the interaction of individuals, groups of people engaging in activities, and most particularly, organizations that are working toward a common goal. Engineers often work in teams where a strong need exists for clear communication. Regardless of how skilled an engineer may be, he or she will not succeed without the ability to effectively inform, persuade, instruct, debate with, and convince other people. The engineer also needs to be able to receive, understand, interpret, and respond to such communication. Listening is a critical element of communication, including within a team framework.

Effective communication in meetings is especially crucial because this may be the only time that the leaders and team members can discuss the project together. Setting an allotted amount of time for the meeting and creating an agenda beforehand can save time and resources. In addition, creating minutes during a meeting and reviewing them at the start of the next meeting is important to keep all group members up to date on what was accomplished in the previous meetings and what issues still need action. It is vital that each team member arrive to the meeting on time (i.e., promptness is an important value) and be prepared with any necessary materials. Teleconferencing, which is becoming more and more popular, may be used. In this case, the engineer should make sure that his or her setting is appropriate and that his or her full attention is devoted to the meeting. A meeting should be fairly structured, allowing ample time for routine activities, as well as an open forum where participants can provide input.

Communication outside of meetings is also vital. Communication between and among team members and leaders should be continuous throughout the project. Team members and leaders should be updated regularly on progress and maintain a constant stream of communication. However, it is important to keep in mind that miscommunication can occur, especially in e-mail or notes, where the correct tone or intention may not be communicated. Team members and leaders should be as clear as possible in their communication methods and be sure to ask for clarification when necessary.

12.8 COMMITMENT TO CHANGE AND INNOVATION IN TEAMWORK

Innovation should not be a reactive process, but one that gives direction. A common sense of purpose should be a fundamental attitude of the team. This will balance the present needs of producing and marketing engineered products and services with the longer-term requirement of research and development (Adair, 2005).

Flexibility is a key quality to innovation. The economic, ecological, and social environment is always changing. A flexible team is capable of responding and conforming to such changes. Communication is vital in these situations, where communication barriers between groups such as researchers, engineers, and the customer should be minimized for the most efficient and profitable response (Adair, 2005).

12.9 MOTIVATION IN TEAMWORK

Motivation influences productivity and is a factor in job satisfaction. Motivation reflects an individual's needs, goals, and values and influences a person's actions. A manager who understands the needs and goals of subordinates will be successful, as he or she will be able to channel the subordinates' actions to meet team or organizational goals. A person's motivation comes from within, but a manager can provide incentives that reflect the subordinates' needs and goals.

While theories of motivation focus on the individual, the concepts apply to teams. If the individuals within a group hold organizational goals to be important, then their internal drive will motivate each team member to achieve the team objectives. The significance of the incentives will influence the degree of productivity, as motivation is not a dichotomous, i.e., on-off, characteristic. The group manager may need to provide incentives that are specific to each member, as each individual in the group will have different needs, goals, and values.

It is vital that group members know what is expected of them and the standards to which they must adhere. High standards give a sense of achievement when attained, which increases the motivation of the group. Group members who constantly receive feedback on their efforts are more likely to develop an internal motivation to complete assignments. If a team of engineers was expected to build a bridge that was to be on budget, on time, and to specifications, then they should be evaluated and informed of their team performance. The team should then be praised for fulfilling or exceeding requirements and should be informed of any shortfall so that they can learn from the experience.

12.10 CREATIVITY IN TEAMWORK

Group leaders need to be creative both to solve problems that invariably arise in fulfilling the objectives of work and to motivate team members to complete their responsibilities. For purposes of this discussion on teamwork, creativity can be defined as the ability to take new and original actions for the purpose of ensuring that the team meets its objectives. Conflict is a specific situation where creative ability is important. Conflicts between team members or conflicts related to team responsibilities often require the group leader to be creative. Very often, the creative solutions to problems yield the best rewards.

Every team member has specific responsibilities, and problems related to each responsibility can be expected. Members must be willing to discipline themselves to use their creative abilities, be receptive to the ideas of others, and work to improve upon other team members' contributions. Critical evaluation should be a step that is separate from the generation of ideas. A positive environment and mutual

> ## CASE STUDY: THE 3M COMPANY
>
> The salesmen at 3M, which started out as a sandpaper-making company, noticed that workers painting two-toned cars in auto plants were having difficulties with the paint colors running together. Richard Drew, a 3M lab technician, soon invented masking tape, 3M's first tape. Soon, Scotch Cellophane Tape was created, originally for industrial packaging. A 3M sales manager, John Borden, created a dispenser with a blade, and Scotch Tape became popular in households around the nation.
>
> The invention and progress of Scotch Tape by the 3M Company is an excellent example of creativity in teamwork. Engineers should constantly keep an eye open for ideas and seek new, innovative solutions. Building upon other team members' ideas is also a vital part of the team process (Adair, 2005).

encouragement are also needed to encourage creative thought, and it is the team leader's responsibility to provide for this environment. Ideas should be subject to rigorous evaluation, so the ability to give and receive criticism effectively should be learned and accepted.

12.11   EVALUATION OF TEAM PERFORMANCE

Ultimately, the evaluation of team performance is based on the extent to which the team met the stated goal and objectives. However, evaluation goes well beyond the single assessment of meeting the objectives. Team performance will excel if both formative and summative evaluations are made (see Chapter 8). Evaluations should be made on individual team members, the team leader, and the team as an entity. Periodic appraisals should be included on the project timeline, which should help motivate team members to meet their responsibilities in a timely manner.

Teams are assembled because the problem requires multiple types of expertise and the workload can be distributed among team members. The latter generally improves efficiency, while the former provides higher-quality work. Ultimately, the team product is evaluated based on the extent to which the project goal and objectives were met; however, the summative evaluation of the product will reflect on the team members, especially the team leader. Evaluation of the product will be based on whether or not the problem was resolved, the timeliness of the effort, and the resources required. For example, if the organizational leader believed that the team required excessive overtime, then justification may be required. A principal focus of evaluation will be on the team leader.

12.11.1   APPRAISAL OF TEAM MEMBERS

Very often, the primary evaluation criterion for a team member is the extent to which the responsibilities/assignments were met in a timely manner and with quality output. This criterion is important because it can reflect on the overall performance of the team. A relatively poor-quality section of the final report can detract from the overall assessment of the project, both the report itself and the team leader. Where a team leader believes that a team member is procrastinating or producing poor-quality output, this should be brought to the attention of the person. An early formative evaluation can result in an attitude change in time to prevent the problem from affecting other team members, such as producing disharmony among team members who must work together.

12.11.2 Appraisal of Team Leader

Since the organization bestows responsibility for the project on the team leader, he or she will receive special credit or blame for the quality of the final product. Criteria used in formative evaluations of team leaders include the planning and organizing of team activities, conflict management, creative problem solving, and sensitivity to organization goals. The organizational leadership will periodically meet with the team leader to receive reports on progress made, the problems that arose, and to receive requests for resources. While the summative evaluation of the team leader's performance may have a greater effect on his or her long-term reputation, the formative evaluations will influence the summative evaluation.

The summative evaluation of the team leader may be conducted by several organizational leaders rather than the one to whom the team leader reported on a regular basis. The extent to which the goal of the project was met will be a principal criterion used to evaluate the team leader. The quality of the product, its timeliness, and the resources required to complete the project are generally secondary criteria used in the summative evaluation of the team leader.

The summative evaluation generally concentrates on two leadership behaviors: structure and consideration. The former addresses issues related to the activities central to functioning of the team. This includes activities such as intergroup communication, the development of subgroups to increase efficiency, the use of resources, and timeliness of reporting. The second factor, consideration, addresses issues related to group personnel. Criteria such as team harmony, morale, turnover, and job satisfaction are used to judge the extent to which the leader successfully managed the personnel.

Each of these evaluation criteria should be considered by the leader both during the project and after completion of the task. Knowing the criteria by which a leader will be evaluated at the start of the group task will generally increase the likelihood of success. For a leader-to-be, appreciating the importance of evaluation criteria, such as the general classes of structure and consideration, and the specific criteria, such as communication and values, can help prepare him or her for success. Also, when the leader-to-be is preparing to move into a position of leadership, he or she must plan for success by addressing these criteria prior to the first group meeting.

12.12 DISCUSSION QUESTIONS

1. What is the difference between intradisciplinary and multidisciplinary teams, and why is it important for engineers to have practical experience in both? Give one example of each type of team that would work on a bridge project.
2. Describe ways that team members and leaders can increase the effectiveness of intradisciplinary and multidisciplinary teams. What are the advantages of being more effective?
3. How can an engineer function effectively as a member of an intradisciplinary team? What opportunities are available to civil engineers?
4. How can an engineer function effectively as a member of a multidisciplinary team? How will taking courses in nonengineering fields like humanities and business help an engineer work in a multidisciplinary team?
5. Describe a major construction project that had been completed in the last decade and one that had been completed over two decades ago. How do these projects differ in terms of teamwork and the number of people involved?
6. Why is it important for the public and public policy officials to be informed about and involved in major construction projects?
7. Describe a situation where there was a communication problem in a team that you were a part of. How did you or your team deal with this problem? Was the solution effective and why?
8. Discuss the relevance of the four factors important to team success to operating a concrete canoe team.
9. What characteristics of teamwork are important to a team of undergraduate civil engineering students solving a group project for a civil engineering class? Discuss.
10. Identify three key characteristics that differ between intradisciplinary and multidisciplinary teams.
11. Identify and discuss three key factors that affect an intradisciplinary team's functioning effectively.
12. Identify and discuss three key factors that affect a multidisciplinary team's functioning effectively.
13. Identify criteria that could be used to evaluate the performance of an intradisciplinary team.
14. Identify criteria that could be used to evaluate the performance of a multidisciplinary team.
15. What factors would be important in organizing an intradisciplinary team? Discuss.
16. What are benefits of time management when conducting research?
17. Identify rationalizations that students may use when trying to justify their poor time management.
18. When involved in group research, can poor time management by one member of the group be considered unethical? Explain.
19. Why is knowledge of time management skills important to the researcher?
20. Identify five factors that contribute to poor time management.

21. What determines whether an item is identified by a researcher as a short-term or long-term goal?
22. Define a timeline. How detailed should a timeline be?
23. Should evaluations of team member performance be just an overall rating (e.g., exceptional, very good, fair, poor) or very specific on items such as quality of work, amicability with team members, etc.? Discuss.

12.13 GROUP ACTIVITIES

1. Create worksheets for formative and summative evaluations of an office for a student chapter. These should be developed for the office itself and not the evaluation of a specific person.
2. Identify a specific student team activity, such as concrete canoe or an Engineers Without Borders project. Develop a detailed plan for the activity and organizing the team.
3. Eight values were identified as being relevant to teamwork. Additional values could be included in the list. Define each value and discuss in detail how each one is important in the activities of a team.

REFERENCE

Adair, J. 2005. *Leadership for innovation: How to organize team creativity and harvest ideas.* Philadelphia: Kopan Page Limited.

13 Attitudes

CHAPTER OBJECTIVES
- Identify attitudes important for civil engineers
- Understand the effect of attitudes on professional practice
- Provide guidelines for improving attitudes
- Discuss the effect of organizational structure on attitude development

---

13.1 INTRODUCTION

For each of the following pairs, select the one attitude that you would want your group leader or project manager to have:

- Self-confidence versus a lack of confidence
- Optimistic versus pessimistic
- Honest versus ethically questionable
- Responsible versus irresponsible

When you consider each of these pairs, the desired attitude is obvious, but this demonstrates that a leader's attitudes are important. How important are these attitudes to success? Very important? Moderately important? Not important? Actually, a person's attitudes are critical to success, and contrary to myth, attitudes can be changed. A person who lacks confidence can develop self-confidence. A "half-empty" pessimist can become a "half-full" optimist. Changing your attitudes so that they are more in line with those generally exhibited by leaders can greatly increase your chances of becoming a leader.

Attitudes, which are very important to a professional, are likely formed by heredity but shaped by education and experiences. An engineering employer certainly views an employee's attitude as a very important characteristic. In this sense, attitude is used as a general term, but attitude can also be applied to specific personal characteristics. For example, a person may seem self-confident, with confidence being the attitude. A person may seem honest because he or she is deemed trustworthy. Curious and creative are other attitudes of importance to professional engineers. A person's success depends greatly on his or her attitudes.

With this background in mind, *attitude* can be defined as a fairly stable, even habitual, disposition that controls a person's thinking, behavior, or tendency toward an action. Both the attitudes of leaders and those of subordinates are important. Employee attitudes affect the overall efficiency of a company, as they influence the actions and decisions of the personnel. The attitudes of a profession, which are influenced by the attitudes of the membership and professional leaders, greatly influence the impact of that profession on society. Thus, discussions about attitudes are important because they are a reflection of the profession, the companies within the profession, and the individuals that make up the profession.

13.2 ATTITUDES AND JOB RESPONSIBILITIES

The attitudes required to meet job responsibilities vary with the position that the person holds and the nature of the work. In all cases, success depends on attitudes. To be successful, a leader will need a different set of attitudes than those required by a successful researcher. A design engineer who works alone will need a different set of attitudes than an engineer involved in teamwork. Some attitudes like self-confidence and honesty are important to all civil engineers regardless of their position. An engineer who lacks an attitude that is important to a position should seek to develop his or her ability reflective of the needed attitude.

A leader (see Chapter 11) has responsibilities including motivating subordinates, decision making, acting as a role model, resolving conflicts, evaluating the performance of individuals and teams, developing business opportunities, and being aware of business trends. The leader will be effective in meeting each of these responsibilities if he or she has the proper attitude. To motivate subordinates requires persuasiveness, fairness, and respect. When involved in making organizational decisions, the leader will need to have a strong commitment and self-confidence. To be an effective role model, a leader will need to be respected and have very high standards; honesty and integrity are also essential to success. When evaluating the performance of employees, fairness and commitment are necessary. A leader who is responsible for developing business for the company needs to be self-confident, have an entrepreneurial attitude, and be persuasive. Most importantly, a person who has aspirations to become a leader will need to develop these attitudes.

A design engineer needs to have a set of attitudes that are somewhat different than those of a leader. Self-confidence is a basis requirement because the engineer will be required to provide initial direction of the tasks. Perseverance will be necessary when any mundane tasks are necessary as part of the design. Integrity is essential to ensure that shortcuts are not taken, such as using someone else's design work. The design engineer will need to be committed to both the design project and the organization.

Civil engineers who work as part of a team (see Chapter 12) will succeed if they are tolerant, fair, considerate of others, respectful, and have good judgment. A lack of tolerance may lead to emotional outbursts such as when a procrastinator on a team fails to complete his or her work in a timely manner. Team members need to be fair in allocating the workload to each team member. The more gifted individuals on a team need to be considerate of those who are less talented than they are. They also need to show respect for other team members. Attitudes are important in teamwork.

Engineers who are involved in research require a quite different set of attitudes. To be successful they need to be confident that they can overcome problems that always arise in research. They must be curious in order to identify areas where new knowledge is needed. Innovative ideas require creativity. Persistence in the face of problems is necessary. Success requires the researcher to be thorough to ensure that the results are adequately verified. Obviously, a researcher should be ethical, as an unethical researcher can unknowingly redirect the work of other researchers to the point where they waste resources following a direction based on faulty work. Researchers need to be optimistic that they will find a solution. While teaching knowledge is important to a researcher, the person's success will heavily depend on his or her attitudes.

13.3   ATTITUDE: COMMITMENT

Each professional should feel bounded to his or her responsibilities. This is commitment. In general terms, *commitment* is the attitude of being responsible, emotionally or intellectually, to some ideal, a course of action, or an issue. Applying this general definition to the professional, commitment is the attitude of boundedness to the values and code of ethics of the profession, to the organizational goals of the employer, to the welfare of subordinates, and to the health, safety, and welfare of the public.

Commitment is multifaceted, multidirectional, and both internal and external to the organization. Just as a professional should have a strong commitment to the employing organization, the employer must reciprocate and have a commitment to the professional and personal goals of the employee. Both the employee and the employer must be committed to responsibilities external to the organization, specifically to the public and the profession. Commitment to sustainability should be important to each person, both as a professional and a citizen.

Being committed to all of one's professional responsibilities should be recognized as beneficial to the individual himself or herself, to the organization, and to society. Commitment should increase one's satisfaction with on-the-job experiences and increase one's efficiency because commitment generates a feeling of ownership to the completion of work assignments. An organization that is committed to the employees will strive to maximize their job satisfaction.

Professionals will not develop an attitude of commitment unless they sense that those to whom they are committed reciprocate with a commitment to them. The professional must feel that the organization at which they are employed will expend the effort to satisfy their personal and professional goals, such as assisting in professional development and ensuring an equitable work environment. The professional must feel that the company is doing its best to provide job security and simultaneously interesting work. Most employees want flexibility in their hours and opportunities for continued growth within the company. Good mentoring will also increase the professional's commitment to the company.

The professional must feel that the profession is also committed to the person's goals. The profession shows this by providing opportunities for professional development, such as offering workshops on the latest technical developments and publications that will enable the individual to keep abreast of the state of the art. The opportunity to participate on committees that lead to advances in technical aspects of the profession goes a long way toward developing the professional's commitment to the profession and the professional society.

13.3.1 Developing an Attitude of Commitment

Individuals often feel committed to many personal entities, e.g., family, a local team, a community group. These feelings of commitment seem to come naturally, yet they are often the result of experiences that have demonstrated a continual reciprocation of a responsibility to personal goals. The same feeling of commitment to an employer does not come automatically on the first day of work. The feeling grows over time when both the individual and the employer continually put forth the effort to meet their responsibilities. The following are a few of the actions that an engineer can take to develop a sense of commitment to the employer:

- Have a positive attitude about organizational goals.
- Do not view your paycheck as the company fulfilling its obligation to you.
- Seek out a mentor within the firm.
- Seek challenging work assignments and complete them with conviction.
- Participate and organize social events within the firm.
- Mentor new employees so that they feel the firm is committed to their personal and professional development.

To paraphrase John F. Kennedy's declaration: Ask not what the company can do for you, but what you can do for the company. Such a perspective will help you develop a sense of loyalty.

This is not to imply that a company does not have responsibilities to the employee. It reciprocates by providing opportunities for lifelong learning, being sensitive to family responsibilities, providing mentoring, and meeting expectations for challenging work assignments.

13.4 ATTITUDE: HIGH EXPECTATIONS

How is success measured? At the end of one's career, a person will not feel successful if the easy pathway was always chosen. If the individual had low expectations for himself or herself, then he or she likely chose the alternative pathway that had a high chance of completion but little professional or personal reward, i.e., little feeling of success. A pathway that was based on high expectations, and possibly a significant likelihood of failure, will likely lead to outcomes that have a greater sense of accomplishment. It is better to choose a career path that has the potential for significant long-term success rather than a path where only small accomplishments are possible. One consequence of always taking the easy path is ultimately a sense of a lack of accomplishment, which can lead to disappointment as well as slower professional advancement.

If someone has low expectations for himself or herself or if a subordinate has low expectations, what can that person do to encourage higher expectations? First, the person must assess why the easy pathway is chosen. Most frequently, a lack of confidence is the constraint. In other cases, the unwillingness to take the risk of failing keeps a person from seeking success at a difficult task. Past failures can cause a person to be reluctant to attempt activities for which they are unsure of success. For these reasons,

- Keep a self-confidence log book in which you document all instances where you feel or act in a nonconfident manner; then record how you should have handled the situation if you had greater self-confidence.
- Do not expect perfection from yourself, as it prevents you from completing tasks, which itself prevents you from experiencing success.
- Do not internalize group failures as being your failure; only accept the portion of failure that you know was your fault.
- Identify your strengths and recognize that these are important to achieving self-confidence.
- Recognize that success is as much the effort as it is the results of the effort, so a strong effort at a task should be viewed positively.
- Take on risky assignments, as these are often a greater potential gain in self-confidence than is the safe assignment.
- Seek out and learn from constructive criticism and recognize that criticism may be warranted.

13.5.3 DEVELOPING SELF-CONFIDENCE IN OTHERS

One task of a leader is to ensure that all subordinates are confident. Many will not initially have self-confidence, so the leader must work to improve the self-confidence of subordinates. Giving praise and open recognition for completed work is important. Encouraging subordinates to be aware of ways of improving self-confidence, such as those in the previous section, is essential. Subordinates who lack self-confidence must recognize that this negative attitude about themselves is actually common and also very correctable.

13.6 ATTITUDE: CURIOSITY

Curiosity can be loosely defined as the desire to learn. This can apply to learning about things related to current needs, e.g., new engineering materials or new design methods, as well as an interest in learning about things that are unrelated to immediate needs, e.g., historical incidents related to changes in technology or novel methods of teaching young children the alphabet. A person's sense of curiosity depends on how frequently he or she has a need-to-know feeling. For example, when a person is in the checkout counter at a grocery store, is he or she thinking of ways that the efficiency of processing customers could be improved? As another example, if a person was playing a game of miniature golf, the person who thinks of ways to make the game more challenging has a sense of curiosity. A professional who thinks about ways of decreasing the rate of rust formation of steel is likely the curious type, unless of course that is the person's job.

Curiosity is an important characteristic of a professional. If the civil engineering profession will need to solve the problems of the future, those in leadership positions will need to be individuals who are generally curious. Curious individuals are generally eager to learn new things, and are therefore attracted to problem-solving situations.

13.6.1 DEVELOPING A SENSE OF CURIOSITY

If a person who is not naturally curious recognizes the importance of enhancing his or her sense of curiosity, the best approach is to set a goal of learning a few basic facts about a variety of topics. This might mean devoting an hour a week in the library looking at books unrelated to professional concepts or past interests. For the engineer, this could include topics related to the arts, a social science, philosophy, a religion different than his or her own, an ancient culture, or some skill like glass blowing. Topics that intersect the civil engineering profession, e.g., business management or environmental law, can also be investigated, but it is more important to approach topics for which an immediate or potential need is not evident.

A questioning attitude is a key ingredient of a curious nature. At every opportunity, a curious person questions either why something is the way it is or how it could be improved. Why doesn't a kitchen toaster have a timer on it to let the meal preparer know exactly when the toast will be ready? What effect would enclosing a storm water detention basin within a grass buffer have on the volume of sediment getting into basin? Developing the habit of asking questions both related and unrelated to one's profession will enhance a person's sense of curiosity.

13.7 ATTITUDE: CREATIVITY

The problems faced by society continually change. Many of the problems are quite complex without obvious solutions. In order for engineers to solve these problems, they must have the creative ability to generate new ideas and analyze alternative solutions. A lack of creativity may mean that all alternative solutions to a problem will not be identified and the potentially best solution may be one of the alternatives not proposed. Due to a lack of creative ability among those responsible for finding a good solution, the problem may not be solved in a way that most efficiently uses both resources and personnel.

13.7.1 DEFINITIONS

The word *create* means to originate, to bring into being, or to produce. It is generally assumed that creating means that the product was the first of something. For example, Rutan created the first airplane that was flown solo around the globe without refueling.

The word *creativity* is applied to individuals who have the ability to create. They have the imaginative powers, i.e., mental processes, and the attitude to recognize important problems and to develop new solutions to solve the problems.

13.7.2 THE CREATIVE PROCESS

The creative process is a way of solving a problem in an original way. Using this process generally leads to a useful solution or product. The scientific method is usually viewed as including the following four steps: observation, hypothesis, experimentation, and induction. The creative problem-solving process can be viewed in four corresponding steps:

1. Recognition: A period of recognizing the problem and questioning the reasons for the problem.
2. Selection: The phase of identifying the most appropriate creativity stimulator to generate potential solutions to the problem.
3. Ideation: The phase of using imagination to generate ideas that may stimulate a solution.
4. Evaluation: The phase of critically evaluating the ideas of the third phase and developing a realistic solution to a problem.

The literature on creative thinking generally concentrates on phase 3, the idea generation phase. However, recognizing the real problem and turning ideas into reasonable solutions are just as critical to problem solving. Asking questions is a central element in all phases of the creative process.

13.7.3 DEVELOPING A CREATIVE ATTITUDE

Unfortunately, creative ability is something that many believe cannot be enhanced; i.e., either you are born with it or you are not! This is a myth. Like learning to play a violin or hitting a baseball, creativity can be learned. While everyone cannot play the violin or hit a baseball equally well—everyone cannot be equally creative, anyone *can* make significant advances with practice. Initially, reading books and articles will give an understanding of the methods and the confidence to try using one's creative powers. Ultimately, it is the successful use of creativity stimulators and the overcoming of creative inhibitors that will enable a person to develop solutions to important problems. Creativity stimulators include brainstorming, brainwriting, and the synectic method (see Appendix B). Creativity inhibitors are many, including the internal feeling that creative thinking is ineffective, an unwillingness to take risks to solve a problem, and the succumbing to discouragement of others.

13.8   ATTITUDE: HONESTY

Honesty is a central value to professional codes of ethics. This is as true in the business world as it is in engineering and medicine. All professions depend on their members to be honest. When a member of any profession acts dishonestly, the act can create doubt over the honesty of other members of the profession. For example, sports talk show hosts questioned the honesty of all basketball referees after one of them was found to be betting on games. For the individual who acts dishonestly, repercussions to his or her career are also likely. The individual suffers, as does his or her employer and others in the profession as well.

Honesty means not to lie, deceive, cheat, steal, defraud, or take unfair advantage of another person or employer. An honest person must be equitable, sincere, and fair in all activities. While a professional must be honest with himself or herself, he or she must be honest with the employer and clients of the employer.

While nearly everyone considers himself or herself to be honest, everyone has the capacity to act dishonestly. A person who is under pressure to succeed at something important to him or her may consider acting dishonestly in order to achieve the goal. The extent to which a person entertains the thought of being dishonest depends on his or her level of moral/ethical maturity. For the ethically mature person, acting dishonestly will, at most, only be a passing thought. For the ethically immature person, the alternative of dishonesty has a much greater chance of being the basis for the decision. The ethically immature person will likely only consider the short-term consequences of the act, most importantly, overcoming the pressure to succeed. The ethically mature person fully recognizes the long-term consequences of being dishonest, which are almost always negative, especially in the feelings that he or she has about himself or herself after reaching ethical maturity.

Dishonesty by an individual can have consequences to all elements of the ethical responsibilities of a professional, which are self, employer, client, the profession, and society. When a dishonest act is exposed, the individual will likely suffer consequences, which may be only a loss of face, but possibly the loss of a job, loss of professional licensure, and even expulsion from the profession. The employer will also be affected by the act. Investigating a potentially dishonest act requires time and resources. A proven act of dishonesty will likely have future consequences for other employees, as company policies will be adopted to prevent such acts in the future. These can reduce individual efficiency as well as cause the employees to be more fearful and have less job satisfaction.

Where the act of dishonesty relates to work for a client of the company, the act may cause the client to be less confident about the company's final product, and thus be less likely to use the company for future projects. The feelings of the client about the company may extend to feelings about the profession as a whole. As an example, politicians have a poor reputation because of dishonest acts by a few.

13.8.1 Self-Enhancement of Ethical Maturity

Reading material on both general and professional ethics appears to be the most fruitful way of enhancing ethical maturity (see Chapter 15). Systematically analyzing professional codes of ethics should be the first step. Using codes of ethics from different professions, review each for the human values they promote, the responsibilities indicated, and the types of activities that they address. Second, review case studies of ethical lapses to understand the types of actions that are viewed by experienced professionals as being unethical. Third, read about specific topics that are common to many cases of unethical practice, including rationalization, self-imposed or external pressures to act unethically, selfishness, and conflicts of interest. Topics like plagiarizing, whistleblowing, and kickbacks also make for interesting reading. Fourth, read and think about the proper ways of resolving ethical dilemmas. After all, people's reputations have been tarnished because of mishandling cases even though they were trying to do the right thing. With all of these avenues of study, the effect will be most pronounced if you discuss what you read with other people. The dialogue should reinforce the lessons that you learn from the readings.

13.9 ATTITUDE: PERSISTENCE

A thirty-five-year-old major league player starts off the baseball season in a 1-for-35 slump. He starts to think that maybe he has lost the ability to get around on fastballs. He starts to think about the possibility of retirement even though at the start of spring training he thought that he was in the best shape of his career. After his latest 0-for-4 performance, he decides to tell the manger to bench him. It seems that the ballplayer is not very persistent. With this attitude, would you expect him to turn around his season and begin to hit well?

Just as persistence is important for the ballplayer, persistence is important to engineers. Persistence refers to a person's tenacity or steadfastness in an undertaking. Engineering designs can be complex with numerous problems arising in all phases of the design. When confronting a problem, the engineer must be resolute in his or her effort to solve the problem. The engineer cannot just accept an easy solution if it would involve significant compromises. He or she must tenaciously seek the best solution. This is persistence!

Being persistent has numerous advantages, as well as some disadvantages. A persistent engineer will likely create opportunities for success, as he or she will likely be successful in finding worthy solutions. Persistence suggests loyalty to the firm's goals, which will also bring recognition.

In addition to advantages, being persistent can be disadvantageous. Excessive persistence may waste time. Team members may not appreciate delays that result from excessive persistence, as it can reflect on their problem-solving ability. Excessive persistence also can lead to a reputation of closed-mindedness. In spite of these disadvantages, persistence is a positive trait.

13.9.1 DEVELOPING AN ATTITUDE OF PERSISTENCE

A person can develop an attitude of persistence by:

- Develop a long-range perspective, where the quick solution may not be a long-term solution.
- Problems frequently arise in research, so get involved in research and always have the positive attitude that a solution can be found.
- Observe cases where giving up after a short period of time failed and other cases where persistence paid off.

13.10 ATTITUDE: PERSUASIVE

Persuasion is an element of communication. The purpose of persuasion is to get others to agree with your position on an issue. This might mean, for example, convincing other design engineers that your choice of a design method is the best alternative in the current project. As a person in charge of personnel for your company, you might want to persuade a sharp young engineer that your company is his or her best job opportunity. These two examples illustrate the need to be persuasive.

Persuasion implies changing another person's thinking such that he or she makes a decision that may have been counter to his or her original line of thought. While a person may act in a way solely because of being in a subordinate position, persuasion generally reflects a change in position or belief primarily because of sound, rational reasoning. A person may be persuaded to change his or her decision in order to receive a reward or compensation. We generally think of persuasion as an oral skill, but an engineer would also benefit from being a persuasive writer, such as when trying to persuade a prospective client through a business proposal letter to award the project to his or her company.

13.10.1 Developing the Ability to Persuade

Persuasion is not an innate skill. In fact, one can develop the skill by understanding and following a few fundamental rules. First, you must be knowledgeable about the subject and the factors that control the person's decision making. For example, being fully versed in the technical aspects of the issue is important, as this inspires the decider to have confidence in your recommendation.

Second, you must be credible. This might include having a credible, respected reputation based on your past experiences. Your credibility may be related to your technical abilities and your reputation for ethical conduct. Past transgressions can reduce your credibility even if they took place more than a decade ago.

Third, you should know the receiver of your message. One of the first rules of interviewing is to research the company before you go to the interview. This rule also holds when you try to persuade another person to adopt your viewpoint. Is he or she rational? Selfish? Emotional? Knowledgeable? Does he or she have the necessary resources? Is he or she impressionable or street-smart? If you know what the person considers important, you can tailor your message appropriately. In a position of leadership, you will learn that one person may value a year-end monetary bonus while another person may value a couple of extra days off. Knowing what people value can help you persuade them to take the less desirable work task.

Fourth, always be positive. This will have an emotional appeal to the message receiver. Your positive attitude will subtly put him or her in a more agreeable mood, which will encourage him or her to make the decision that is in line with your position.

Fifth, you should be organized, especially when it comes to the ways that people will benefit if they adopt your position. If you identify the benefits, your position will be strong and counterarguments will seem trivial.

13.11 ATTITUDE: OPTIMISM

- There is no solution!
- We are just wasting our time trying to solve this problem!
- There is no way I can pass this test!

These are statements of pessimism. Contrast them with the following:

- I am sure that a solution to the problem exists!
- If I stay positive, I know that I can pass the test!

These are statements of optimism. Which of the two attitudes do you believe will lead to success?

Optimism is the belief that good outcomes will occur. Optimists have hope that a positive outcome is more likely than a negative outcome. While the pessimist dwells on what can potentially go wrong, the optimist focuses on the potential positive outcomes of a situation.

How might an optimistic outlook be beneficial in one's professional life? First, the optimist will likely try harder to solve a problem, while a pessimistic person will likely get discouraged early on and thus give up trying to solve the problem. This is true in engineering management, research, or leadership. Optimistic managers believe that they will be awarded contracts, and so they put more effort in tasks such as writing proposals. The pessimistic manager does not put much effort into developing a business plan because he or she does not expect that the effort will be rewarded. Similarly, when faced with what may seem an insurmountable problem, the optimistic researcher believes that a solution to the problem will be found with just a little more effort. What attitude will the pessimist adopt?

Optimism also benefits job satisfaction. The optimistic person who is surrounded by pessimists may be unhappy in his or her position. If the general feeling is that the likelihood of success in professional endeavors is low, a sense of discouragement will likely arise. Conversely, an office filled with optimistic engineers will be an office where individual leadership arises, honesty is never questioned, and the tougher assignments are valued.

13.11.1 Evaluating Your PO Attitude

The first step to move away from pessimism is to do a self-evaluation and rate yourself on a linear scale of pessimism to optimism. Begin by listing five (or more) recent activities in which you were involved. This could include taking a test in a tough course, preparing a paper for a liberal arts or humanities course, going on a date with someone new, playing in a playoff game for an intramural sport, and traveling a long distance to a new place. Try to select activities that are dissimilar to each other. Now think back to the time just prior to the event. Did you feel that you had little chance of doing well on the test, i.e., "That professor always gives unreasonable tests"? Or maybe you had the thought, "I will likely not have a good time on the date because the person will think I am too nerdy." Develop a scale to rate each of the activities. For example, you might use −1 for a pessimistic view and +1 for an optimistic view, with a rating of 0 to indicate neither an optimistic or pessimistic view, i.e., neutral. Now rate each of the activities using your pessimistic–optimistic (PO) scale.

Obviously, a numerical value based on the sum of the individual scores is not scientifically valid. However, on a broad scale, it may suggest a tendency toward optimism or pessimism. Obviously, if you put a +1 by each activity, then you may not need to worry about your level of optimism. A negative sum, however, may suggest the need for an attitude change.

13.11.2 DEVELOPING AN ATTITUDE OF OPTIMISM

To overcome a tendency toward pessimism, you must first acknowledge that changing this attitude would have many benefits. It could lead to greater academic success, a better social life, improved friendships, and better motivating and problem-solving skills. Your reputation as a leader will also be enhanced.

Developing an optimistic attitude will require you to have a greater awareness of your feelings. When you face a test, a new date, or a confrontation with a friend, you must evaluate your thoughts. If they are primarily pessimistic, evaluate each reason and turn it around so that it is expressed as a positive. Go from "she will think I am a nerd" to "she seems to be the type that highly ranks intelligence." Go from " I will do poorly on the test because the professor always gives hard tests" to " I have a 3.2 GPA, so I know that I can do as well on this test as I have on most of the others." This turn-around-an-attitude plan requires concentrated and continuous thinking about the attitude problem.

As a follow-up to this thought process, you should maintain a PO diary. In this diary, you should record all instances of pessimistic thoughts and what optimistic arguments you should have used to move away from the pessimism. Then record the outcome of the event: the date went well or you passed the test with a better than initially expected grade.

The PO diary should also be used to record statements by friends who have optimistic or pessimistic tones. Following each entry, record your thoughts about the person who made the statement. This is not a statement about the person overall, but about your immediate perception about the person specifically due to the statement. For example, a friend declares: "This team project will be a failure because my team members are losers." This is obviously a pessimistic statement, so you may think: "He will not succeed because he doesn't recognize that he has significant control over success." The pessimism keeps the friend from recognizing that he has a role in the project success. In the PO diary, the student should enter something to the effect, "My friend's pessimism will be as much a factor in failure as will the poor effort of his teammates." What lessons can be learned from the observation, the thought, and the PO diary entry?

You should solidify your commitment to having a positive attitude after several weeks of making entries in your PO diary. You should summarize your assessments of a friend about whom you have a number of entries. If he or she seems to have a negative score on your PO scale, discuss the pessimistic attitude with him or her to help him or her overcome the problem and recognize the benefit of having an optimistic attitude.

13.12 SELF-EVALUATION OF ATTITUDE

Improving your attitude to achieve success requires an attitude self-evaluation. Self-evaluation is an activity that focuses on your current status and the type of person that you want to be. If you want to be a leader, you will need to develop attitudes that will allow you to be a leader. If you wish to be a successful researcher, then your plan should seek to develop attitudes important to those in research. If you are uncertain about your future, then the goal should be to enhance attitudes that are important universally. For example, attitudes of honesty and creativity are necessary for success in all professional endeavors.

A simple attitude of self-evaluation can be made by listing all attitudes that you believe are important. Then develop a rating plan that can be used to identify your attitude strengths and weaknesses. For those attitudes about which you feel positive, try to assess the factors that contributed to success in this area. For those attitudes that you feel are your weaknesses, identify the origin of the weakness. Then for each attitude, identify actions that you sincerely plan to take to improve yourself. This may involve any of the following: (1) reading about the topic, (2) joining a self-help group such as Toastmasters, (3) maintaining a personal attitude log in which you document successes and failures related to each attitude, (4) seeking help from a person who has expertise related to attitude about which you have trouble, (5) working with a friend who has similar attitude problems to overcome, or (6) volunteering for positions or activities that will enable you to gain relevant experience.

13.13   ATTITUDE CHANGE

A person entering professional life has attitudes that have been developed since childhood. For example, a person may have a pro-environmental attitude that negatively views land development. An anti-nuclear-power attitude may have developed after seeing a TV report on Chernobyl. The difficulties in cleaning up the Alaskan shoreline following the *Exxon Valdez* accident may have produced a no-more-development attitude toward oil production of the North Slope. The formation of such attitudes may have been developed based on narrowly focused inputs, but they still may be strongly felt.

Advertisements for consumer products and political candidates, to name a few, are intended to be attitude-changing messages. The ten-second news clip of candidate X stating something controversial is shown as part of candidate Y's political ad in order to change the viewer's attitude about X so that the viewer will vote for Y. All advertisements are intended to change the viewer or listener's attitude. Given the amount of money that advertisers are willing to spend on ads for radio and TV attests to the effectiveness of this form of "education."

Education also changes attitudes. Education about the effectiveness of different patterns of cluster housing can change the student's attitude about land development. Education about the safety of nuclear power can change a person's attitude toward the expansion of nuclear power for generating electricity. Of course, the effect of education on attitude change depends on the source of the message, the extent to which the message appeals to the receiver, and the characteristics of the recipient of the message.

13.13.1   EFFECT OF THE MESSENGER

Three factors are principal determinants of the effectiveness of the messenger: credibility, power, and likeability. Credibility implies a worthiness of belief or confidence in what the messenger says. It implies that past messages have proven to be reliable.

With respect to changing attitude, power reflects the capacity of the messenger to exercise control over the attitudes of the receiver or the ability to influence the attitude. Likeability has two indicators. First, when the recipient of the message has an inclination toward the message sender, likeability is high. Likeability also implies an equivalency of the characteristics between the two.

Credibility depends on the messenger's believability. If he or she is known to be an honest, knowledgeable person, then the messenger is viewed as being credible. The accuracy of past messages distributed by the messenger influences the level of credibility. In cases where the messenger is unfamiliar to the receiver of the message, the credibility is largely influenced by the credibility of the position the messenger holds. For example, following Hurricane Katrina, the FEMA spokesperson had little credibility because of the inaccuracy of some of his statements.

The power of the messenger is connected to the credibility of the messenger. However, independently it depends on the position of the person, the image of confidence conveyed, and the human relationships with the recipients of the message. Power and control or authority are related, as the position of the messenger often carries with it a level of authority or the ability to dictate. Positional power is often highly associated with the position of the person in the organization. However, personal power can be increased by doing personal favors for the recipient, increasing one's expertise, or increasing the recipient's dependency, such as with the control of resources.

While credibility and power are understandably important determinants of the willingness to change an attitude, likeability is also a very influential factor, possibly beyond the extent that it should influence a person's willingness to change. Should a good personality really be influential in changing another person's attitude? Likeability depends on the similarity of the recipient with the messenger and any friendship that develops. If the mayor of a city is liked by his or her constituents, then they may be more likely to follow his or her dictates about responding to a natural disaster, e.g., to evacuate the city due to an approaching hurricane.

13.13.2 Effect of the Message

Your attitude about an issue may have been influenced by too little information or background knowledge. For example, a biased report on the effects of logging may influence a person to believe that all human-initiated deforestation is wrong. Presenters of messages intend to change people's attitudes, often employing somewhat devious means. For example, they may present a one-sided analysis or possibly present very weak arguments to counter the thrust of their message. The message may also refer to a celebrity who favors the position. The message receiver is more likely to favor the message if he or she respects or likes the celebrity to whom the message refers. Creating fear in the mind of the message receiver is another tactic used by those wanting to change a person's attitude. Advertisements about safety belt use may have induced those who feared having an accident and subsequent disability to wear their seat belt; however, the fear factor was not effective on everyone, which made seat belts laws necessary. A message receiver must analyze each message to identify biased or fear-inducing statements intended to get him or her to act in a particular way. The analysis is intended to understand whether or not the messenger used devious means in an attempt to change attitudes about a position.

13.13.3 THE MESSAGE RECEIVER

A primary determinant of attitude change is the needs and goals of the message receiver. If the person recognizes the need to have more self-confidence and sets that as a personal goal, then efforts to increase self-confidence are likely to be successful. The strength of the need will influence the long-term adherence to the goal.

13.13.4 CHANGING ONE'S OWN ATTITUDE

A self-assessment is the first step in developing a more mature set of attitudes. For example, if a person now recognizes the importance of being more curious or the advantages of being more self-confident, then action is warranted. The following actions can be used to change one's own attitude:

- Identify the benefits of changing.
- Develop a plan to accomplish the change.
- Get information on ways to accomplish change.
- Set reasonable goals and a time plan for achieving them.

Regardless of the desired change, these actions should lead to accomplishing the objective.

13.14 CREATING AN ATTITUDE-SENSITIVE ORGANIZATIONAL STRUCTURE

Given that attitudes are important both personally and for professional advancement, organizations need structures that foster attitude development. Organizational practices need to be open in order to maintain high levels of employee job satisfaction. Organizations need to establish policies and practices that represent a favorable environment for attitudes that are supportive of practice. Possible elements that are sensitive to attitude development are:

- *Develop an effective mentoring program.* Mentoring is a very effective practice for positive interpersonal relationships. The mentor can emphasize the attitudes important to success within the organization and those necessary for assuming leadership positions.
- *Ensure communication lines are open.* A lack of communication can result in misunderstandings, which are a primary factor in the adoption of negative attitudes.
- *Increase task accomplishment.* Workers who have a sense of accomplishment generally adopt positive attitudes. Thus, the organization needs to provide the support and resources necessary to ensure continued task completion.

- *Develop a balanced reward system.* Awarding rewards for positive attitudes may be difficult because of the lack of a strong link between rewards and performance. However, while not strong, the positive performance-reward link should be used to advantage. The reward system can be oriented to either individuals or teams. The latter has the advantage of reducing jealousy between individuals, especially when it is organization-wide and uses objective criteria.

These are just a few of the practices that play a positive role in team building. An organization takes on the attitudes of its leaders, so leaders need to exhibit the attitudes that the organization promotes. The proper use of power is important. Power should not be used for controlling or pejorative purposes. Practices such as the proper use of power demonstrate the effect that leadership can have on attitude development.

13.15 DISCUSSION QUESTIONS

1. List attitudes that would be important to a BS degree holder who serves as a project manager. Rank these in order of importance and briefly discuss why each is important.
2. List attitudes that would be important to the president of a medium-sized civil engineering firm. Rank these in order of importance and briefly discuss why each is important.
3. Explain why curiosity is important to project managers of small engineering firms.
4. Explain why curiosity is an important attitude in conducting research experiments performed as an undergraduate.
5. Explain why honesty is an important attitude for the president of a mid-sized engineering firm.
6. Explain why self-confidence is important to an entry-level engineer.
7. Develop a case study that demonstrates the importance of self-confidence to a project manager who must address a community group about the siting of a waste treatment facility in their neighborhood.
8. Give examples that would illustrate how a project manager for a small engineering firm would demonstrate commitment to the projects that he or she manages.
9. Give examples that would illustrate how a graduate student would demonstrate having high expectations for his or her graduate research project.
10. A vice president of a mid-sized engineering company is preparing a proposal to submit to a client on the design and layout of a multibuilding office complex. This would include parking, drainage and storm water management, and security, as well as the buildings. Analyze this case on the basis of the attitudes of high expectations, creativity, and public safety.
11. Identify the steps of the experimental process (see Section 4.5) and discuss one attitude that would be important at each step.
12. Identify and discuss the attitudes associated with lifelong learning. Analyze the attitudes to show how they can help an engineer effectively learn after completing his or her formal education.

13. Identify and discuss the attitudes associated with a dishonest person. Contrast them with the attitudes of an honest person.
14. Analyze the process of whistleblowing and discuss the attitudes relevant within the process.

13.16 GROUP ACTIVITIES

1. Create an organizational structure for mentoring within a company in which the mentor stresses the importance of attitudes needed to complete engineering design projects.
2. Review biographical information on a historical figure in engineering (e.g., George Goethals, John Wesley Powell, Washington Roebling) and evaluate his or her attitudes that enabled him or her to be a leader. How did his or her attitudes lead to success?
3. Identify the general steps of the engineering design process. Assume that a design will be completed by a team of engineers, some located in different countries who collaborate via the Internet. Discuss the attitudes important for the team to work effectively.

REFERENCE

Zarnoth, P., and Sniezek, J. A. 1997. The social influence of confidence in group decision making. *Journal of Experimental Social Psychology* 3:345–66.

14 Lifelong Learning

CHAPTER OBJECTIVES
- Discuss ways of lifelong learning
- Identify skills and attitudes important to lifelong learning
- Introduce approaches to lifelong learning
- Present a framework for developing a lifelong learning plan

14.1   INTRODUCTION

Through the mid twentieth century, changes in technology were generational rather than occurring annually. The education provided by a BS degree, along with knowledge and experience accumulated over one's career, was then sufficient. Times have changed! Now technology changes over very short periods of time. Research results are introduced into design standards in a matter of a few years rather than a few decades. New materials have and are developed. Our understanding of physical phenomena such as earthquake loadings and effects has improved rapidly. Computer advances have made it possible to incorporate greater complexity into design methods. Just consider that those currently approaching retirement probably used the slide rule for computations in their undergraduate years. Just think how computational technology has changed over their professional careers. Experimental evidence identifies more accurate effects of design inputs. Assessments of failures have identified weaknesses in design procedures or construction methods, thus precipitating change. Changes in public laws have required changes in design practices. A changing natural environment and a greater understanding of the need to conserve natural resources have also added to the knowledge requirements of the practicing civil engineer. All of these changes have required the practicing engineer to continue to learn well beyond completion of his or her baccalaureate degree. Just as society wants medical doctors to know the latest in health care methods, society expects the civil engineer to know the most recent knowledge about advances in civil engineering materials, models, and design methods.

14.2   LIFELONG LEARNING: A DEFINITION

The ASCE *Body of Knowledge* defines *lifelong learning* "as the ability to acquire knowledge, understanding, or skill throughout one's life" (ASCE, 149). Lifelong learning is not just an ability, but the sincere interest in maintaining professional competency. Lifelong learning goes beyond technical matters and includes career advancement

---

REASONS TO PRACTICE LIFELONG LEARNING

- Professional advancement
- Update out-of-date knowledge
- Forestall obsolescence
- Personal fulfillment
- Broader knowledge base
- Deeper understanding of a specialization
- Greater confidence in completing assignments
- Learn new design methods, models, and materials

---

issues, leadership skills, areas of liberal learning, ethical and social responsibilities, improved historical understanding of the effects of technology, legal aspects of engineering practice, cultural issues, and even personal development. As global practice becomes the norm, the civil engineer will need a much broader knowledge than has been sufficient in the past. As each subdiscipline of civil engineering becomes more complex, the civil engineer will need to assimilate these advances and simultaneously appreciate new knowledge of other subdisciplines of civil engineering.

A professional who practices lifelong learning will be better able to fulfill his or her responsibilities to the employer, clients, profession, and society. The engineer who is self-motivated to stay abreast of changes in knowledge will be more likely to experience professional advancement within a company. Personal fulfillment and intellectual stimulation are two additional benefits of lifelong learning. The lifelong learner will likely have more self-confidence to provide professional service. If the lifelong learning goes beyond the purely technical knowledge, the achievement of a broader knowledge base will lead to greater opportunities for challenging assignments.

14.3 VALUES RELEVANT TO LIFELONG LEARNING

Lifelong learning is based on values. Each civil engineer has a responsibility to maintain professional competency, and as the profession changes rapidly over time, lifelong learning is required for the engineer to meet his or her responsibilities to the public and the employer. Values relevant to lifelong learning include

- *Knowledge*: Understanding gained through study
- *Industriousness*: The quality of being diligently active in study
- *Accountability*: Answerable for obligations to maintain competency
- *Credibility*: Worthiness of belief; reliable
- *Competence*: The state of being capable

The relevance of knowledge and its importance to a design engineer are obvious. Failing to know the best design method and being uncertain of the design parameters can lead to an unnecessarily risky design. Adequate knowledge is more than just knowing how to put data into a computer program that uses the latest knowledge. Having adequate knowledge means knowing the technical methods on which the program is based, as well as the proper way to interpret the computer output. The required knowledge is likely more than what is taught in a college course or a workshop. Adequate knowledge is based on relevant experience gained after the in-class learning.

Self-study is not something that should be approached with the attitude "I'll study the material when I need to know it." After all, at that point, the time available to learn the material may not be sufficient. Thus, the engineer should be curious and industrious and view learning as a continuing responsibility, and not study at a need-to-know pace. If lifelong learning is a responsibility, then maintaining competency depends on self-accountability. Without the knowledge gained from lifelong learning, an engineer's credibility is lessened, as he or she cannot be relied upon to provide competent design solutions.

14.4 SKILLS AND ATTITUDES IMPORTANT TO LIFELONG LEARNING

Imagine working from sunrise to sunset and having to negotiate traffic jams both on the way to the office and then on the return trip home. Now imagine having to devote an hour or two of your evening to studying the latest material on design methods. This picture is not one that will likely generate enthusiasm for meeting a responsibility to maintain professional competency through lifelong learning. To achieve success in fulfilling this responsibility will require special skills and a professional attitude.

What skills are required to meet the responsibility to maintain professional competency? Learning skills used as an undergraduate or in other formal education are the starting point. The undergraduate should value perseverance and be disciplined. The lifelong learner will require these, as many have difficulty studying without the planning that accompanies formal education. Organization ability is another essential skill. This includes organizing both one's schedule, i.e., time management, and one's strategies for learning. True learning is more than just being exposed to facts. Instead, the facts must be organized in a logical manner. To be learned effectively in a way that the material can be recalled and used requires the facts to be placed within a more general framework that provides a more meaningful context.

14.4.1 THE PROCESS OF SELF-STUDY

Organization is the key to developing a self-study program. In many ways, the process of self-study follows the steps used in organizing any activity. Steps basic to organizing any activity include

1. Identify the goal and objectives of the learning responsibility.
2. Establish the major tasks and any subtasks that will be needed to complete the tasks.
3. Identify resources, especially time, that will be needed to complete each task; developing a timeline is a good way of specifying time requirements.
4. Develop evaluation criteria needed for self-assessment of learning and indicate on the timeline of step 3 when self-assessments will be made.

Even though self-study is important to a person's success, it is still a difficult endeavor. Allocating the time required to organize a self-study activity can make the process more efficient and enjoyable and increase one's chance for success.

14.4.2   Time Management

Time management is a primary skill for meeting self-study objectives. It increases learning efficiency, helps one achieve learning goals in a reasonable time frame, and increases personal satisfaction. Also, one success at effective time management will provide you with the confidence to approach more challenging self-study topics. In time management, distinguishing between controllable and uncontrollable times is important, as you have few options with uncontrollable time (e.g., time required to commute, sleep, eat). With respect to self-study planning, in identifying controllable time you will have to consider personal responsibilities and other professional obligations. Managing controllable time is where you can apply the principles of effective time management.

If lifelong learning is viewed in terms of efficiency, then wasted time means less knowledge gained per unit of effort expended on learning. Time wasters then contribute to inefficiency. If you can eliminate, even minimize, time wasters from your schedule, your productivity of self-study will increase significantly. The first step is to identify those time wasters that affect your schedule. These may include requiring perfection, saying yes to requests that are unimportant, and waiting for a large block of time before beginning to start your self-study program. Attempting to make the first draft of a self-study plan perfect will delay your progress and very likely reduce your confidence in achieving success at other self-study activities. Similarly, taking on responsibilities that deflect your attention from your goal is counterproductive. One of your first steps in developing a self-study plan is to determine the times that you will likely have to devote to self-study. Will the times be short periods of a half-hour to an hour, or larger blocks of time, say three hours on a weekend morning? This will influence how you organize your self-study syllabus. If you only have short time periods available, you will need to divide the self-study work into small activities that likely have a chance of being completed in the allotted time. Procrastination and the inability to eliminate time wasters can undermine your entire self-study program.

Some transition time is probably necessary, but if the times become too long or too frequent, then they become time wasters and keep you from fulfilling

---

DEFINITIONS: TIME MANAGEMENT

- *Controllable time*: Time that you are free to schedule activities.
- *Time wasted*: Time spent on unimportant activities, specifically those that are not relevant to your goals and objectives.
- *Transition time*: Time between significant parts of your time schedule.
- *Procrastination*: A needless postponement of an activity that should be completed at the present time.

your self-study objectives. One very effective way to accomplish a task is to divide it into smaller tasks, ones that will seem more manageable. This was part of the second step of the process of organization. One problem that can result from creating many small tasks is that you fail to control the transition times. If you waste the transition times, the overall productivity will decline. Dividing a one-hour task into two thirty-minute minitasks may create a transition time that is really unnecessary and may encourage time wasting. Thus, transition time control is important.

14.4.3 PROCRASTINATION

Procrastination is a major contributor to inefficiency. The procrastinator often argues, "Procrastination is actually a good thing in my case because I work better under pressure!" Study after study has shown this to be untrue. Procrastination does not lead to better work, and it is especially damaging when the work is part of a team effort, because one person's procrastination can limit the effectiveness of another person's work when that person depends on the procrastinator's work to complete the task. Overcoming procrastination requires effort and a sincere understanding that procrastination is a serious fault.

14.4.4 IMPORTANT ATTITUDES

Curiosity and perseverance are two attitudes important for effective self-study. Curiosity will provide the motivation needed to approach a new topic, especially when the subject is not an immediate on-the-job need. If your current job assignment requires knowledge of a new design method or a new material, then fulfilling your responsibility may be sufficient motivation to learn about the topic. However, when the subject is not an immediate job responsibility, effective self-study will depend on a sense of curiosity to provide the motivation. Perseverance is the attitude of steadfastness in the pursuit of a goal, which would be learning a subject through self-study. Perseverance is necessary to maintain the motivation through to the completion of the self-study task.

An innovative attitude is also helpful to achieve self-study goals. Materials in professional journals and trade magazines often omit important details, which is necessary because space is limited. Therefore, a person trying to learn new technical material will often need to synthesize the missing steps. This may require some creative thought or the ability to answer the question, "How do I get from this equation to the next equation?" Innovative thought will be necessary to answer this.

14.5 SELF-STUDY PLANNING

What made it possible for you to learn so much about so many topics while you were an undergraduate student? Obviously, you can point to the long hours spent studying. But even before you started a class, many things were happening that made it possible for you to learn so much. Specifically, the course instructor did a lot of planning and organizing from which you benefited. This included the following:

- The instructor decided which topics were important and how much time would be required to adequately cover each topic.
- The instructor prepared a syllabus that identified the order of the topics, the time to be spent on each topic, and assignments to help you assess your progress.
- The instructor selected a text book(s) and other reference materials.
- The instructor prepared lectures to identify relevant background, theory, and applications for each topic.
- The instructor prepared supplemental handouts for each topic.

Without this planning, it is doubtful that you would have learned so much in such a short period of time.

The focus on the instructor's preparation is relevant to lifelong learning because your efforts at self-study will require you to mimic the planning and organizing of your former instructors. If your efforts in maintaining competency are to be successful, you will need to decide on the topics about which you need to or should learn more, the sequence to follow, and the resources that will help you learn the material. Planning will require you to establish a time schedule and set up a place where you will do the self-study. Only after planning and organizing will your efforts at studying be successful. Learning how to learn before you try your hand at self-study is essential to your success.

14.6   APPROACHES TO LIFELONG LEARNING

As technological, legal, and cultural changes have taken place, an industry for continuing professional development has developed around the traditional methods. Advanced degrees, trade magazines, attendance at local section meetings of professional societies, and professional journals have been the traditional means of maintaining professional competency. Printed documents were popular because they could be transported with the person and read when time permitted. For those who needed personal interaction to learn, formal education for an advanced degree was the most common traditional approach. As changes associated with the practice of engineering have increased at a fast pace, short courses and workshops have become a more common pathway to continuing professional development. These courses are attractive because the learning is compressed into a short time frame, generally two to five days, and an expert is there to answer questions. Distance learning facilities have eliminated much of the travel problems related to these workshops. Advances in technology have benefited those who pursue lifelong learning. The array of lifelong learning options has and will continue to expand as the demand increases and technology advances.

14.7   DEVELOPING A LEARNING PLAN

It is quite likely that every engineer will need to develop a self-study program at some point during his or her career. Indicators that you should begin a self-study program might be one of the following:

- You read an article in a trade magazine that addresses an issue about which you should be familiar, but you wonder how the design could be done.
- People at the office or those whom you meet at local professional meetings seem to know more about an issue than you.
- A regional work group is formed to study a new problem and you are not asked to be a member even though you have worked in the area of interest for some time.
- You are using a new computer program to do design work and you know how to get the input data into the program, but you are not sure about the procedure on which the program is based.

Given the rate at which technology and design methods currently change, starting a self-study program as soon as you believe that you are lacking in knowledge is important.

Planning for the future is not an exercise characterized by certainty. This is especially true of young engineers, and especially pre-BS students. However, having a plan for any of life's endeavors is still a good idea. The plan must allow for change, such as relocations or family problems. The benefits of an education can be many, including having goals, having an understanding of the importance of continued professional development, recognizing that long-term success will depend on keeping abreast of state-of-the-art knowledge, and integrating learning into one's career.

One approach to developing an educational plan is to create a timeline that lists major objectives, such as completing the undergraduate degree, getting summer internships that provide valuable experience, taking the FE examination, identifying a mentor while you are an undergraduate, enrolling in graduate school on a part-time or full-time basis, getting involved in civil engineering research, selecting a full-time employer who will provide challenging work assignments, taking the PE examination, and approaching managerial responsibilities. By developing a timeline of these career milestones, the need for continued educational development should be evident. For example, if managerial responsibilities are a long-term goal, then leadership experience should be one of your short-term objectives. Just identifying a position in management as a ten-year objective and not including a leadership development activity at some earlier point on the timeline could lengthen the time before you achieve your goal of being part of management. Creating the timeline will make it evident that short-term experiences are necessary to achieve long-term goals.

Options for gaining additional knowledge are many, and the number of options is likely to increase in the future. The following are a few of the options:

- Enroll in a graduate program.
- If you have completed an advanced degree, periodically enroll in new courses that cover advances in the field.
- Review professional journals and trade magazines.
- Attend symposia or local section meetings.
- Develop a local section committee to study an issue of current concern and distribute a report on the issue that can be used by others to know the latest information on the issue.

- Seek out experts and discuss recent advances and trends on a topic.
- Scan the Internet for discussions of issues that you face.

It is important that you spend some portion of your workweek on lifelong learning. It is easy to put lifelong learning off to some time in the future, but because technology and design methods are advancing at such a fast pace, delaying learning will likely hinder your advancement.

14.8 EVALUATION OF A SELF-STUDY ACTIVITY

Consider courses that were part of your undergraduate program, including the non-engineering courses. If you were to evaluate the courses, criteria for assessing the extent to which you learned the material might include

- I could see how individual topics related to the general theme of the course.
- I thought that I could apply the material.
- I understood what concepts were important as they were taught.
- In subsequent courses where I needed the material, I felt prepared to learn the new material.
- I recognized the relevance of the material to my future career.

Obviously, other criteria could be stated, but criteria such as these would be useful in deciding whether or not the effort expended to learn the material was worthwhile. Note that the grade received in the course was not a criterion used to evaluate the depth of learning.

Evaluation that follows one activity is very useful for ensuring success at future activities. Therefore, evaluation is an important component of self-study activities. Self-study evaluation provides numerous benefits, most notably increasing learning efficiency of future self-study activities, greater knowledge gain, and greater interest in taking on future self-study activities. The following are questions that could be used to perform a self-evaluation of a self-study activity:

- Were all self-study objectives met? If not, why?
- Were the setting and facilities for your self-study activity conducive to learning? If not, what were the major distractions?
- Was the time allocated for each session well managed? If not, what changes in time management should be made?
- Was the self-study activity well planned? If not, what changes should be made in planning future self-study activities?
- Would the activity have been more efficient if periodic evaluations had been made? If so, how frequently should evaluations be made during the course of a self-study activity?

Devoting time to evaluating your initial attempts at self-study can increase your confidence in self-directed learning.

14.9 DISCUSSION QUESTIONS

1. Define lifelong learning and discuss topics to which this refers.
2. Identify general reasons that lifelong learning is important to a medical doctor. Then associate each reason to a person who practices civil engineering.
3. Explain why knowledge of other cultures might be important to a civil engineer. Identify ways that a practicing engineer could learn about different cultures.
4. Explain why knowledge of the history of technology might be important to a practicing structural engineer. List sources that discuss the technology that would be relevant to civil engineering.
5. Obtain a recent copy of *Civil Engineering*. Select an article that discusses a project about which you are interested. While reading the article, make a written list of concepts with which you are familiar and a second list that includes topics about which you would need more knowledge in order to fully understand the article. What do these lists suggest?
6. Obtain an announcement for a workshop/short course that is offered by ASCE. Evaluate it on the basis of the cost per hour and compare this with tuition per hour at a local university. Examine the course content relative to that of a preparatory course offered as part of a BSCE program. Does the short-course announcement indicate the experience of the instructor? If so, what aspect of the instructor's experience is important to you?
7. Outline a self-directed learning plan that could be used to investigate the potential value of nanomaterials to a civil engineer.
8. Outline a self-directed learning plan that could be used to investigate the potential value of sensors to a civil engineer with a specialty in bridge pier design.
9. Provide reasons why knowledge of electrical engineering, such as lasers and sensors, might be important to a practicing civil engineer.
10. Identify two professional topics that would not be considered technical subjects that you envision yourself pursuing as part of your self-study plan. This could include subjects like professional ethics, the history of technology, or oral communication. Outline specific subject matter related to each topic that you believe should be addressed. Identify ways that you could advance your knowledge in both of the topics.
11. From your current academic position (i.e., undergraduate student, BSCE degree holder, graduate student, etc.) develop a continuing education plan that will cover the next five years. While specifics about the future may not be possible, identify your options.
12. Approach a college professor and discuss with him or her the steps that he or she uses in organizing and planning to teach a new course. Then use the tasks identified to plan a self-study program on leadership skills needed in civil engineering.
13. Identify major changes in computational tools (e.g., slide rule, hand calculator, mainframe computer, microcomputer, PC, etc.) along with the approximate dates of their use. Discuss why lifelong learning was necessary for the civil engineer to stay abreast of changes in computational technology.

14. Discuss ways that a person could evaluate his or her success at self-study. Make a list of criteria that could be used.

15. Make a list of criteria that could be used to evaluate a two-month long summer trip around Europe. Then transform each criterion into one that would be relevant to evaluating an educational activity, such as a short course.

16. Select a course that you took at least two years ago. Since that time, identify instances where you had to use the subject matter and evaluate your successes or failures at applying the material. What factors contributed to your successes or failures?

17. Self-study learning as part of the body of knowledge refers to personal or professional growth that involves changes in oneself. Discuss how the self-study of new subject matter can change a person.

18. Discuss the following statement: A person learns significantly only when the topics studied are perceived as being important to the learner's personal goals. How is this idea relevant to lifelong learning?

14.10 TEAM PROJECTS

1. As a project, select a topic about which you have very little knowledge or experience. Such topics might include nanomaterials, macrobiotics, green roofs, desalinization, debris flows, or engineering in a specific ancient culture. Conduct an investigation of the topic such that you learn enough to write a three-page paper on its potential relevance to civil engineering. Then evaluate the process that you used to learn about the topic and assess how you could improve the process for future self-directed learning endeavors.

2. Find an announcement for a two- or three-day workshop on a subject that might be of interest to you. Review the list of topics to be presented at the workshop. Then obtain textbooks where each individual topic is discussed. Discuss why attendance at the workshop might be of value, beyond self-study with the textbooks.

3. For one of the four following topics, obtain a recently published textbook that discusses the topic in detail:
   - *System simulation*: The simulation process; random number generation; model development; analyzes simulation results.
   - *Tertiary wastewater treatment*: Chemical coagulations; dissolved organic adsorption; nitrogen and phosphorous removal; distillation.
   - *Decision making involving risk*: Alternative criteria for making decisions; expected value decisions; expected value decision making; decision making by simulation; incorporation risk into decision making.
   - *Restoring wetlands*: Flood pulsing; passive versus active restoration; natural vegetation; the restoration process.

   For the topic chosen, develop a syllabus that you could use for self-study of the topic. Indicate the specific topics to be covered and the expected time required to cover each part of the lesson. Do not select a topic you have already studied. Identify your self-study objectives and several books that you could use as reference material. Topics other than those above could be used.

4. Obtain course syllabi for at least a half dozen academic courses; try to include a couple from outside of engineering. Detailed syllabi are more useful than brief syllabi. Analyze each syllabus and identify the fundamental components included. Then discuss with a faculty member the steps that he or she uses in planning and organizing lectures, assignments, and tests for presenting a new subject. Based on these activities, develop a self-study plan for a subject such as ecological sustainability, ethical issues in engineering practice, or handling risk in engineering projects. Present your self-study plan as a formal document.

REFERENCE

ASCE. 2008. *Civil engineering body of knowledge for the 21st century.* Reston, VA: ASCE Press.

15 Professional and Ethical Responsibilities

CHAPTER OBJECTIVES
- Introduce codes of ethics and the underlying values
- Define and interpret the term *professionalism*
- Show the relevance of the code of ethics in resolving ethical conflicts
- Present a strategy for addressing ethical problems
- Discuss rationalization

15.1 INTRODUCTION

Consider the following case studies.

15.1.1 CASE STUDY

Craig Westcott is responsible for awarding subcontracts for ABC, Inc., which is a large construction company. An employer of XYZ Engineers asks Craig out to lunch to discuss XYZ's qualifications with the hope of securing future subcontracts. Craig agrees and XYZ picks up the check.

Should this be considered a business lunch and an acceptable practice, or is it a bribe that places Craig in a conflict-of-interest position?

Business lunches are generally considered acceptable, but if Craig awards a subcontract to XYZ Engineers following the luncheon, could other subcontractors insinuate that Craig was bought? Could another subcontractor argue that Craig violated the phrase in the code of ethics that states that engineers should be honest and impartial?

15.1.2 CASE STUDY

Rachel Eastman is a PE who works for Traf-Surv, Inc., a small company that does traffic surveys. Rachel does the fieldwork and writes the reports for clients that detail the results of traffic surveys. Recent surveys have shown that new traffic control equipment is not needed, which is not the news that the client wants to hear. Rachel's boss, who is not an engineer, suggests that Rachel should show loyalty to the client, as the client would be much happier if Traf-Surv showed that new traffic control equipment is necessary. Rachel interprets his wording as a directive to change the data to show results favorable to the client.

Should Rachel show loyalty to the client? Should she go to the local newspaper and report her boss? Should she succumb to the dictates of her office manager?

If these are all of the alternatives that Rachel identifies, then it is evident that she needs mentoring about how to handle ethical dilemmas. While engineers have obligations and responsibilities to clients, she would not be meeting her responsibilities to society and the profession if she acted unethically. Going public with revelations about conduct of questionable ethics is rarely a proper decision. The third alternative is unethical. Rachel would have developed a better list of alternatives if she had an educational foundation in applied ethics.

15.1.3 CASE STUDY

A large contractor is building a regional shopping mall on a site that is adjacent to a river. Instead of paying to have the excess building materials hauled away, the construction site superintendent has the workers dump much of the trash into the river. Trevor White does not like this practice. In addition to the legal issues, Trevor recognizes the environmental consequences of the practice. His pleas to the superintendent are ignored.

Is this an ethical issue? If so, how should Trevor handle the situation?

Professional codes of ethics include statements about sustainability and others about public welfare. Do such statements apply only to nonrenewable natural resources, or are they applicable to the health of the river? Codes of ethics are not lists of do's and don'ts, so they do not include statements about dumping trash into rivers. Does this imply that dumping is not an ethical issue about which engineers should be concerned? Several elements of a professional code of ethics do apply to this case study.

These three case studies indicate the difficulty with ethical decision making. The decisions seem obvious, but in an actual case, the proper action to take is rarely obvious. Very often, even professionals would not agree on the best alternative. However, failure to make a good decision about an ethical dilemma can damage a person's career. Knowing what the civil engineering profession expects of its members is important.

15.1.4 EXAMPLES OF ETHICAL ISSUES

The three case studies in the previous sections introduce three types of activities where values and ethical decisions play a role. The number of such activities are limitless. New technologies always create new value dilemmas. For example, should civil engineers use a new material that requires considerable use of a nonrenewable resource? The growing concern for the environment and sustainability introduces new considerations in engineering design. For example, should a potential danger to an endangered species cause a contractor to shelve a project? This latter case is difficult to answer because it may not be certain that the project would impact the species. Engineering risk and uncertainty (see Chapter 7) are often associated with value and ethical issues, which must be considered in both design and engineering decision making. The practice of engineering has business requirements, so ethical issues discussed in business management courses are also relevant to engineering. The point is that values and ethics are important and a major consideration in the practice of civil engineering.

While value and ethical issues are numerous, a listing of types of ethical issues that civil engineers confront may suggest the breadth of the concern. Some of these issues relate to business practices, while others relate to new technologies, risk, public safety, and the reputation of the engineering profession. The following is a short list of potential situations that involve ethical dilemmas:

• Data fabrication	• Inadequate supervision of subordinates
• Data falsification	• Plagiarism of another's design
• Bid rigging	• Accepting excessive gratuities
• Making kickbacks	• Unauthorized reading of emails
• Dual publishing	• Signing documents without checking
• Ghost authorship	• Not reporting environmental pollution
• Résumé padding	• Failure to account for risk in design
• Disclose military secrets	• Unnecessary cost overruns
• Contingency fees	• Leaking information to the press
• Falsify travel expenses	• Failure to inspect design work
• Copyright infringement	• Failure to maintain competency

Each of these ethical dilemmas involves competing values, and a civil engineer must first recognize all values involved and know the proper method for handling such cases.

15.2   VALUES AND VALUE SYSTEMS

A *value* is a principle, character trait, standard, or quality considered worthwhile or desirable. This would include pleasure, knowledge, freedom, equality, and love, just to list a few. Things like money and new cars may be valued, but they are not values. The importance of values varies with the individual and with time. Freedom is an especially critical value to someone who lives under a dictatorship, while it is given far less consideration by someone who lives in a free society. Some people have a thirst for knowledge, while others emphasize pleasure as a value, which then motivates their behavior. That people assign different weights to different values has important implications in professional life, such as causing intraoffice strife.

The values that are of primary importance in one's personal life may be quite different than the values that are important in one's professional life. Values such as life, happiness, pleasure, and cleanliness are important in one's personal life. When a person enters a profession, the personal values are still important, but he or she must incorporate an additional set of values into his or her decision making. Loyalty, efficiency, and knowledge become very important in one's role within a firm and a profession like civil engineering. As a professional who has value responsibilities to society, values such as public health and safety, respect, and honesty take on a primary role in fulfilling the professional's ethical obligations to society. Honor and the freedom to practice are values of special importance in the relationship between a professional and the profession. Entering a profession means recognizing a broad array of values relevant to conduct.

Loyalty is considered to be an important value in professional life. Unfortunately, like most values, we cannot classify loyalty as being either a good value or a bad value. The quality of the value depends on its role in value decision making. If a decision situation arises where a professional is considering divulging company secrets, then loyalty would suggest that the professional refrain. Conversely, if a professional finds out that company managers are including falsified résumés in proposals, loyalty to the firm would require the professional not to act on the information; however, loyalty to society and the engineering profession would require action to stop the practice. In the case of divulging secrets, loyalty pushes the individual toward the correct decision. In the second case, loyalty may have some positive career implications, but it forces the professional to act disloyally to both the clients and society. This evaluation of the value of loyalty indicates that values may influence decisions, but we cannot create lists of positive and negative values. The context is also important to decision making.

15.3 CHARACTERISTICS OF A PROFESSIONAL

When we hear the word *professional*, we may think of a professional athlete or perhaps even a hired killer. After all, the broadest definition of the word *professional* is one who is employed in an activity because of a need of livelihood. According to this definition, almost all people who work could be classed as professionals. Plumbers, grave diggers, used car salespersons, and chimney sweeps are engaged in specific activities as a source of livelihood. It is not that grave diggers and football players are *not* professionals; they certainly are needed citizens who have skills that are used in a service to society. But certain activities are accorded the status of professions; for example, doctors, members of the clergy, and teachers belong to professions. These professions are usually thought to require a higher level of knowledge and formal education than others who might satisfy a dictionary definition of a professional because they serve society in special ways. It is in this sense that we want to classify engineers as professionals; while engineers engage in specific activities as a source of livelihood, their work carries special responsibilities to society that accords them the status of a profession. These responsibilities will be briefly outlined here.

In his discussion of professional development, Dougherty (1961, 3) defined a professional as

> one who uses specialized knowledge and skill in the solution of problems that cannot be standardized. He is actuated by a service motive; he works in a relation of confidence, and observes an acceptable code of ethical conduct.

It must be emphasized that the term *professional* does not apply only to someone who has specialized knowledge and skill. Other important dimensions of professionalism exist. In his definition, Dougherty (1961) recognized the multifaceted nature of professionalism, and he concluded

> Professionalism is a way of thinking and living rather than an accumulation of knowledge and power. Knowledge and power are essential and when actuated by the professional spirit they produce the leaders and torch-bearers.

What are the fundamental dimensions of professionalism that are necessary for an individual to be considered a professional? The National Society of Professional Engineers (NSPE, 1976) has identified five characteristics of a professional engineer, which also apply to those in the sciences:

1. One who possesses a body of distinctive technical knowledge and art gained by education, research, and experience.
2. One who recognizes a service motive to society in vital and honorable activities.
3. One who believes in standards of conduct, such as represented by ethical rules.
4. One who supports a legal status of admission to the profession. The most common forms are registration to practice engineering and graduation from an accredited engineering curriculum.
5. One who has pride in the profession and a desire to promote technical knowledge and professional ideals.

To these five characteristics, Dougherty's definition suggests a sixth factor that we could add as a fundamental characteristic, even though it may be inherent in the foregoing five characteristics:

6. One who works in a relation of confidence to his employer, the public, and all who use his or her works.

Considerable time, education, and professional practice are required to obtain the specialized knowledge and skill that characterize a professional. The professional, in contrast to the artisan or technician, is capable of synthesizing past experience and fundamental principles in solving new problems. Such problem solving requires independent thought and a motivation that is not self-serving. The motivation must stem from a dedication to the service of both society and the profession; otherwise, the knowledge and skill do not maximize human welfare. Because of the interaction of a professional with society and the advantage that the professional has through his or her specialized knowledge and skill, it is important for a professional to apply sound moral judgment and to be able and willing to translate this moral judgment into principled professional conduct.

15.4   CODES OF ETHICS: OBJECTIVES

The Ten Commandments, the golden rule, and the pledges of Boy Scouts and Girl Scouts are codes of conduct. They are intended to be guides of living and for interacting with people. One version of the golden rule is: Do unto others, as you want others to do unto you. This would apply to children ("I won't hit you because I don't want you to hit me") as well as adults ("I will return things that I borrow from you because I want you to return things that you borrow from me"). While it has applications to professional practice, it is certainly not adequate by itself. It is generally not sufficient as a guide for professional conduct.

In place of such guidelines, professionals adopt codes of conduct or codes of ethics. Most professional societies have codes of ethics, which serve a variety of purposes. First, codes are an outward indication of the values considered important by the profession, as well as the object of the profession's attention with respect to values. Second, codes serve as a model for professional action and potentially serve to counteract unethical conduct. Third, codes of ethics that stress human and societal values demonstrate that professional decisions should not be based solely on technical and economic decision criteria. Fourth, codes stress value obligations and responsibilities to those outside of the firm, thus broadening the decision-making perspective of management. Fifth, they bring attention to the need for professionals to avoid conflicts of interest.

15.5   RESPONSIBILITIES INHERENT TO CODES OF ETHICS

Codes of ethics mention those to whom engineers are responsible, but they also identify the nature of the responsibilities. Engineers are responsible for public health, safety, and welfare, as well as the environment. Codes indicate that engineers should avoid conflicts of interest, be honest, maintain professional competency, not compete unfairly with others, and only make objective and truthful statements. Meeting these responsibilities will provide a professional environment throughout the civil engineering profession.

Knowledge and its advancement are important to the civil engineering profession. It is inherent to professional codes in a number of ways. First, the first fundamental principle indicates that engineers should use "their knowledge and skills for the advancement of human welfare." Second, the seventh fundamental canon of the ASCE Code of Ethics states that "engineers shall continue their professional development throughout their careers." This implies that engineers must maintain competency by keeping abreast of new knowledge in their chosen specialties. As a guideline to practice, the ASCE code specifically states how maintaining professional competency can be achieved.

15.6   VALUE CONFLICTS

Each day we make numerous decisions, from the trivial (what to wear, what TV show to watch, what flavor of ice cream to buy) to the important (to get married, to select a graduate school). Decisions about these actions are often very dependent on our value preferences. But how many people think about and understand the process that they use to make these decisions? For example, does a person critically analyze competing alternatives when making a decision to cheat? Would using a systematic process to make a decision lead to better decisions? Probably not when it comes to selecting an ice cream flavor. When a more important decision needs to be made, a process would be beneficial. Decisions on conflicts that involve professional ethics and values certainly fall in the category of important. As the importance of a decision increases, the quality of the process that the person uses to make the decision increases the likelihood that the best decision will be made.

Civil engineers make business decisions on a regular basis, and many of these decisions involve value criteria such as honesty, fairness, loyalty, promise keeping, and confidentiality. Making an incorrect value decision can be detrimental to one's career, so it is important to recognize value issues and the values that are in conflict. Consider the following dilemmas and the competing values:

- Officers of an engineering company participate in a kickback scheme (honesty, fairness, and integrity versus selfishness and disrespect).
- An engineer does not provide adequate supervision of subordinates who are involved in design (accountability, dependability, and duty versus personal happiness and a lack of prudence).
- An engineer charges time to a contract on which he or she did not work (truth, accountability, dependability, and duty versus personal happiness and a lack of integrity).
- A company does not provide an outside consultant with the work promised when the company used his name, résumé, and reputation to get a contract (honesty, truth, and fairness versus efficiency, a lack of accountability, and a lack of trustworthiness).
- An engineer uses computer code that he developed for his previous employer at his new job (fairness, accountability, and confidentiality versus a lack of loyalty and irresponsibility).

In each of these cases, values conflict. The person making the decision must weigh the values along with other factors, such as economics, reputation, and efficiency. In many cases where values conflict, legal issues must also be considered.

Consider the case of a person who is under pressure to complete a report that involves data collection, analysis, and synthesis. The person may be facing pressures to provide timely reporting, produce positive results to support a particular position, or internal pressure to complete the task so that the individual can move on to other tasks. The individual considers fabricating the data to save time and would produce the desired results. The value conflict is obvious, honesty versus personal pleasure.

What factors would an individual consider in making the decision to fabricate or not to fabricate? Most likely, the principal decision factor will be the likelihood of getting caught. If the individual believes that the likelihood of getting caught is small, then he or she will be unlikely to consider alternative decisions in which honesty is part of the value foundation. The individual will most likely not identify the best course of action.

In the case of the person considering data fabrication, important, unselfish goals, such as gaining knowledge that would come from the data collection and analyses, are not identified. Realistic alternatives that would provide some results, such a scaling back the extent of data collected, are not considered. Certainly, decision criteria such as honesty and accuracy are given very little weight compared to criteria such as expediency and satisfying the stakeholder who wants the results. Failure to properly identify the goals, decision criteria, and alternatives is very likely to lead to faulty decision making. This is just as true in the case of value-based decisions as it is in technical decisions.

15.7 VALUE DECISION MAKING

While a more complete presentation of the value decision-making process is needed, it is worthwhile to identify the important elements. First, it is important to identify the real problem, the goal that will lead to a solution, the decision criteria used to make a decision, the values in conflict, and a preliminary list of alternative courses of action. Second, planning is also important, and this includes gathering pertinent information and developing a complete specification of alternatives. Third, the alternatives should be evaluated using appropriate decision criteria and the best alternative selected.

For less complex ethical dilemmas, such as classroom cheating, the array of both decision criteria and the number of alternatives is considerably less than for a case of whistleblowing that involves considerable technical detail. Understanding the proper application of the principles will lead to better decisions regardless of the complexity of the case.

Decision making is a topic that is central to a wide variety of disciplines. Discussions of decision making are relevant to a diverse array of subjects, including economics and business management, statistics, engineering, medical science, and sociology. Investigators in each of these diverse disciplines have developed procedures for making decisions, which consist of steps that, when followed, will lead to some optimal decision. While the procedures vary in the number of steps, the sequence of certain actions, and the degree of objectivity in the decision rules, they have many commonalities. The quality of the decision scheme is not dependent on the number of steps or the complexity of the process. Instead, the quality depends on the ease with which the process can be applied such that the important elements are not omitted.

While value decision making has obvious differences from engineering, statistical, or economic decision making, it has similarities in that good decision making involves executing a well-deserved process without disregarding important steps. In engineering problem solving, the decision maker must have a sound understanding of the technical problem and the theoretical or empirical method that is most appropriate for the design. In value problem solving, the decision maker must know the value criteria being used on the competing sides of the value conflict. In engineering problem solving, the engineer must develop a solution methodology, which includes identifying all possible alternatives for completing the design. In value problem solving, the decision maker must identify all possible alternatives for resolving the conflict. In engineering problem solving, the decision maker must apply the theoretical or empirical methods to each of the alternative solutions to decide the one that is most appropriate. In value problem solving, each solution alternative must be evaluated using the value criteria to select the optional solution.

In summary, the resolution of value conflicts has much in common with technical problem solving. They both require a thorough analysis of the problem, and failure in any of the early stages of problem solving can result in a nonoptimal solution in the decision stage.

15.8 ETHICAL MATURITY

The word *maturity* is most often applied to measure a teen's level of social skills. The statement "He or she is mature for his or her age" generally reflects the person's ability to interact with other people. The word *maturity* is also applied to the level of development of young children. It's also not uncommon to hear an athletic coach discuss a player's level of athletic maturity, e.g., he was sent down to the minor leagues because he lacked the maturity to show up on time for team meetings. These examples suggest that maturity has many dimensions.

With respect to the general topic of professionalism, a professional's level of ethical maturity is of primary interest. Ethical maturity is the aspect of professionalism that corresponds to the moral maturity in an individual's personal life. Maturity is the condition of being fully developed in a specific capacity. The growth to maturity is a process rather than a dichotomous classification. A teen may not be considered fully mature, yet he or she may not be considered immature. Maturity is a process with a number of stages.

The ethical maturity of a professional is important because it reflects how the individual approaches a value dilemma. Professionals have value obligations to many, including the employer, the client, the profession, and society, as well as to himself or herself. Professionals also have value responsibilities to sustainability and the environment. An ethically mature person recognizes the importance of all of these value obligations and properly weighs each of them when making a decision that involves conflicting values. A person who recognizes the ethical obligations to himself or herself and gives more weight to these obligations at the expense of professional value obligations is considered ethically immature. The interval between ethical maturity and immaturity is a continuous scale, one where the professional recognizes at least some value obligations but may not properly weigh all of the obligations. The individual gains maturity only after he or she is able and freely elects to make value decisions properly.

A professional gains ethical maturity through observation and lifelong learning. Evaluating a person's level of ethical maturity is not easy. To reach ethical maturity, an individual must appreciate values, fully understand the concepts that underlie codes of ethics, know the procedure for handling value conflicts, and be able to properly weigh competing values. The person must view civil engineering as a profession, not just as a business. Believing in the importance of the environment and sustainability is also an indication of ethical maturity. Knowing these central issues and being able to apply them is necessary to develop ethical maturity.

15.9 RESOLVING ETHICAL CONFLICTS

The most important element in solving ethical conflicts within the workplace is to distinguish between internal and external appeal. *Internal appeal* refers to actions taken within the firm to solve an ethical dilemma. *External appeal* refers to actions with respect to parties outside the firm. Except under unusual circumstances, all internal options should be considered in attempting to resolve an ethical conflict before seeking an external solution. The internal options are separated into three levels: individual preparation, communication with the supervisor, and initiation of the formal internal appeals process. It is important to follow the appeals process in the sequence shown in Table 15.1. As the first step, it is to the advantage of the employee to develop accurate records of the facts and details of the conflict, familiarize himself or herself with the appeals process for resolving ethical conflicts within the company, and identify alternative courses of action. At this point, it is also extremely important for the individual to know exactly what he or she expects any appeal to accomplish. It is inadequate just to state that a problem may exist; ideally, the employee can propose a solution that can resolve the problem. The individual must recognize the moral rights of the firm, the profession, and society instead of just showing concern for the effects on himself or herself.

Once the individual has studied and documented the facts and formulated a plan for the appeal, the matter should be informally discussed with the immediate supervisor. In most cases, problems can be resolved at this level, which is to the benefit of all parties involved. In some cases, a more formal appeal to the immediate supervisor is necessary. This appeal is usually in the form of a memorandum that clearly states the facts, the individual's concerns, and the actions that would be necessary to resolve the concerns. It should be evident that good communication skills are important in resolving ethical conflicts because a failure to clearly state the concerns, whether orally or in writing, may hinder the resolution of the problem. Appealing the problem to higher levels before completing the appeal to the immediate supervisor is viewed negatively by all involved and decreases the likelihood of a favorable resolution. Finally, before continuing the appeal beyond the immediate supervisor, an individual should inform the supervisor in writing of the intention to continue the appeal to higher authorities.

The process of appealing an ethical conflict within the firm is usually quite similar to the interaction with the immediate supervisor. Formal steps should follow informal discussions, and steps within the process should not be bypassed. If the internal appeals process is completed without resolution of the conflict, the individual should

---

TABLE 15.1
Procedure for Solving Ethical Conflicts

I. Internal appeal options

    A. Begin individual preparation

        1. Maintain a record of events and details

        2. Examine the firm's internal appeals process

        3. Know federal and state laws

        4. Identify alternative courses of action

        5. Specify the outcome that you expect the appeal to accomplish

    B. Communicate with your immediate supervisor

        1. Initiate an informal discussion

        2. Make a formal appeal

        3. Indicate that you intend to begin the firm's internal process of appeal

    C. Initiate an appeal through the internal chain of command

        1. Maintain a formal communication appeal

        2. Formally inform the company that you intend to pursue an external solution

II. External appeal options

    A. Undertake personal options

        1. Engage legal counsel

        2. Contact a professional society

    B. Communicate with the client

    C. Contact the media

---

formally notify the company that he or she intends to continue with an external review of the problem.

Before expressing to the client the possibility of making concerns public, a lawyer or professional society should be consulted. Lawyers can identify courses of action and pitfalls in external appeal. Whereas lawyers understand legal constraints and consequences, they often lack the technical expertise necessary to evaluate the technical adequacy of arguments. Professional societies have such technical expertise available and may be willing to informally discuss the case and make recommendations; however, they may not have complete understanding of the legal implications.

After engaging legal counsel or obtaining advice from a professional society, the individual may elect to approach the client. In many cases, the client lacks the expertise to judge the accuracy of the claims and will rely on the advice of the firm in evaluating the concerns; however, if the issues involved in the conflict can be accurately communicated, the client may pressure the firm to resolve the issue internally. Alternatively, the client may expend the resources necessary to obtain an unbiased review of the matter. While it is extremely rare that it will be necessary to carry the appeals process to the client, the effects on one's professional career can be detrimental if the results fail to validate the arguments.

The last resort in resolving ethical conflicts is public disclosure. The benefits and costs in terms of both economic criteria and nonquantifiable value goals should

---

CASE STUDY

Lynne, a CE graduate with two years of experience, works for a CE firm that does traffic surveys. A company takes a contract with a manufacturer of traffic control equipment. To market its products, it hires the firm that employs Lynne. After several studies for the manufacturer, all of which showed that new control devices were not necessary, Lynne's project manager, a non-CE, told Lynne that it would be better if her analyses showed that new equipment was needed at these sites. Lynne felt that the manager's choice of words was suggesting that Lynne should falsify the data to show that new equipment would be needed. What should Lynne do? How would her activities differ if the manager specifically told her to falsify the data?

If Lynne is ethically mature, she would immediately respond to the manager that falsifying the data would be a violation of the code of ethics, as it is dishonest and her employer's interests and the public would not be ethically served. Lynne should also make notes on the encounter with the manager, with all of the details recorded. The notes should be kept in a secure place away from the office. While her rebuttal to the manager should be sufficient to prevent future suggestions of unethical behavior, she should be sensitive to the manager's future wording to sense any statement that would suggest unethical action.

---

be assessed rigorously before the media is contacted. It is not in the best interest of society to make public disclosure prior to seeking an internal resolution. The only case where public disclosure is acceptable prior to following the appeals process (Table 15.1) is when immediate danger to the public exists.

15.10 CHARACTERISTICS OF UNETHICAL CONDUCT

We are all familiar with cheating, a general term that is one example of immoral conduct. Almost all of us have observed cheating in the classroom, and now it is not uncommon to hear about cheating in grade schools. *Cheating* can be loosely defined as acts meant to deceive, a dishonest act, or the fraudulent acquisition of another's property. Other specific examples of moral misconduct include plagiarism, padding a résumé, and the falsification or fabrication of data.

It is of interest to analyze these examples of moral misconduct to identify their general characteristics. First, moral misconduct involves values that are in conflict. In the case of cheating, the cheater must weigh honesty against pleasure, among other values that are involved. Second, the individual making the decision has a value system, which encompasses the individual's knowledge of values, the weights applied to the values, and the process followed in making a decision. Very often, the person is ethically immature, and his or her underdeveloped value system allows for the conduct. Third, rationalization (see Section 15.13) is common to almost all cases of moral misconduct, and those who are guilty generally rationalize their actions in

---

CASE STUDY: PROPOSAL PADDING

Bowman and Associates, Inc., is a young firm that is struggling to attract sufficient work to keep all of its professional staff paid. Most of the work is in the area of model development, especially with computers and geographic information systems in the environmental field. In an effort to get a contract from the EPA, Bob Bowman solicits résumés from two very respected professionals, one a retired government engineer and the other a professor at the state university. Bob tells them he will hire them as consultants if he gets the EPA contract on which he used their résumés. However, he knows that the budget could not realistically absorb their consulting fees, so he has no intention of using their services if Bowman and Associates receives the contract.

What values are important to Bob Bowman? What is your assessment of Bowman's ethical maturity? How does Bowman justify his conduct? What are the potential long-term consequences of this practice? Do you believe that he will repeat this practice in the future? If you worked for Bowman and Associates and you were aware of Bob's decision, what action would you take?

---

order to avoid both feeling guilty and changing their value system so that they will not repeat the misconduct. Fourth, the individual lacks the foresight to recognize the potential long-term consequences of the act. Only the short-term benefits are considered in the decision. Fifth, studies suggest that unethical acts are not one-time events, which indicates that the person has likely committed one or more unethical acts in the past. If the act was not discovered, then the person did not feel a sense of failure and probably did not reevaluate his or her value system.

Knowledge of these general characteristics of moral misconduct can provide a professional with the basis for confronting unethical conduct in professional settings. To properly approach ethical decision making, it is important for the professional (1) to have knowledge of values, both human and societal values; (2) to recognize the multifaceted nature of a professional's ethical obligations; (3) to know that value decision making involves the weighing of values, and that the weights an individual applies to the conflicting values reflects his or her value system; (4) to know what the profession expects of its members concerning values and value obligations; and (5) to know the procedure for properly handling value conflicts.

15.11   WHISTLEBLOWING

Whistleblowing is the act of reporting on unethical or illegal conduct within a firm to someone outside of the firm in an effort to discourage the firm from continuing the activity. The whistleblower takes information that the firm would not want publicly disclosed and reports it to an external oversight board or, in many cases, the public media. Whistleblowing could also be applied to the reporting of unethical conduct to someone within the firm, but in a way that bypasses some levels of the chain of command.

Whistleblowing is a major professional action—an act that can have long-term career consequences. It can also have significant implications to public safety. For example, consider the case of a state highway employee with many years of bridge inspection experience who believes a certain bridge is unsafe. He reports it to his supervisors, but they do not believe that the bridge is unsafe and decide to do nothing. The inspector believes the bridge is not safe. Should he blow the whistle to the local newspaper? If he does, he will likely lose his job. If he doesn't and the bridge fails, he could feel guilty for the rest of his life if the failure results in death. If he blows the whistle and an outside investigator does not substantiate the finding of an unsafe bridge, then his professional reputation could be severely damaged. To blow or to swallow the whistle is not an easy decision to make. The problem is complicated when the danger is imminent, with little time to obtain an independent verification of the problem.

15.12 RATIONALIZATION

A procrastinator frequently uses the same excuses to delay getting started. For example, the following excuse is common: "I will do a better job if I wait until just before it is due." In falsifying an income tax statement, a person might use the following excuse as justification to himself or herself: "The IRS takes enough of my income." Such excuses are called rationalizations. The process of making excuses to justify an action that the person knows is not right is generally referred to as *rationalization.*

Rationalization is important because it is often a critical element in making unethical decisions. By rationalizing when confronted by an ethical conflict, an individual will make a self-serving decision and simultaneously avoid feeling guilty about his or her action. It is a part of the individual's value system and is a reflection of that individual's level of ethical maturity.

A formal definition of rationalization is the mental process of falsifying the motivation underlying one's action so that he or she will not regret the action.

Generally, the motivation is to achieve some personal gain at the expense of others. In the case of the income tax falsifier, he or she benefits financially at the expense of the public. In the case of the procrastinator, he or she benefits in the present time at the expense of the future. Both the procrastinator and the income tax falsifier develop their excuse, i.e., the rationalization, to avoid feeling guilty and to avoid changing their value systems, which lean toward being selfish.

In summary, individuals rationalize for four reasons:

1. To avoid feeling guilty, especially in the short term
2. To justify the action in their own mind
3. To avoid changing their value system toward a more altruistic moral stance
4. To try to convince others that the extenuating circumstances are more significant than the misdeed

It is important to understand the concept of rationalization because it is very common in cases of wrongdoing. It is just as important for someone to recognize it when he or she is rationalizing his or her own conduct as well as to recognize it in the conduct of others.

15.13   DISCUSSION QUESTIONS

1. When a code of ethics refers to public welfare, how should an engineer interpret this responsibility?
2. Engineers have principal ethical responsibilities to five entities. List them.
3. Explain the ethical responsibility of an engineer to his or her employer.
4. Explain the ethical responsibility of an engineering company to the local community.
5. Provide an argument for and an argument against the idea that a professional code of ethics that indicates a principal ethical responsibility to public health, safety, and welfare applies to the environment, not just human beings.
6. Discuss the circumstances under which an external appeal can be made without completing an internal appeal of a perceived ethical problem.
7. Give three examples of conflict of interest in professional situations.
8. Does the ASCE Code of Ethics apply to preservation of ecological diversity? Explain your position.
9. It is generally considered unethical not to give subordinates adequate supervision when they do design work. What factors might be relevant in defining *adequate supervision*?
10. Would it be a conflict of interest for a college faculty member who also works for a private civil engineering firm to hire one of his students on a part-time basis? Discuss the issue from both viewpoints.
11. Discuss the ethical responsibilities that a design engineer has to the environment.
12. What does the phrase "one who recognizes a service motive to society" imply?
13. If a business lunch is considered an acceptable gratuity, would a dinner at an expensive restaurant be acceptable? Where is the line between acceptable and unacceptable?
14. An engineer takes his wife along on a business trip and upon his return includes some of her expenses on his travel reimbursement request. Identify elements of the ASCE Code of Ethics that are relevant to this situation.
15. A senior civil engineering student is preparing his résumé for distribution to potential employers. On the résumé, he inflates his responsibilities for his summer job, suggests that he was student chapter officer when he was not, and inflates his GPA for the last two years. What are the value issues involved?
16. An engineer inflates his travel refund request to bill the employer for more than he actually spent. Use a code of ethics to analyze this practice.
17. An engineer works all week on a project for client A. At the end of the week, the office manager tells the engineer that the account for client A has already been spent and that if she wants to be paid she will need to charge her time to the account for client B. Apply standards of professional responsibility to determine her appropriate course of action.

18. An employer inflates the résumé of an employee when submitting a proposal for work. The employee finds out about the inflated résumé. Identify the competing value issues, the possible courses of action available to the engineer, and the decision that the engineer should make.

19. A manufacturing company discharges wastes into a nearby river. According to law, the engineer is responsible for testing for certain pollutants and if the levels in the stream exceed standards, then the engineer is required to stop further discharges and immediately report the testing results to the state EPA. In one instance, the engineer is pressured by the plant manager not to immediately report the results. What should the engineer do? Explain the decision from an ethical standpoint, not a legal standpoint.

20. What rationalizations might an employee use for falsifying his time sheet?

21. Discuss ways that a person can enhance his or her own ethical maturity.

22. Discuss the pros and cons of whistleblowing.

23. Analyze whistleblowing from the perspective of a professional code of ethics.

15.14 GROUP ACTIVITIES

1. Develop a single professional code of ethics that could be used by a company that employs engineers, accountants, and lawyers.

2. Develop a script for a short TV show that shows an engineer confronted by an ethical problem (e.g., a colleague is using a computer program to do design work when the colleague does not really understand that ethical details that underlie competing ethical responsibilities contribute to the conflict) and how the matter is resolved.

3. Obtain a copy of the video *Testing Water … and Ethics*. Identify the values in conflict and discuss the ethical maturation of Jim Duffy over the course of the video case study.

REFERENCES

ASCE code of ethics. 2008. https://www.asce.org/inside/codeofethics.cfm

Dougherty, N. W. 1961. Methods of accomplishing professional development. *Trans. ASCE* 126(V):1–6.

NSPE. 1976. *A guide for developing courses in engineering professionalism.* Washington, DC: National Society of Professional Engineers. Publ. No. 2010. *Civil engineering body of knowledge for the 21st century.* Reston, VA: ASCE Press.

Appendix A: Communication

APPENDIX OBJECTIVES
- To provide information relative to the structure of written reports
- To discuss methods of graphical analysis
- To provide guidelines for making oral presentation

A.1 FORMAL REPORTS

If an engineer has frequent writing responsibilities, then he or she will want and need to know how to write efficiently. Time spent unnecessarily revising drafts of a report is time lost. This is not implying time spent revising written material is time lost, as it is doubtful that any first draft is good enough to be the final document. Being organized will keep the number of rewritings to a minimum. The following sections identify a few factors that contribute to organized writing.

A.1.1 OUTLINING A REPORT

Experienced writers know that outlining is the best way to get started. Outlining forces the writer to focus on the important elements and prevents the writer from trying to use a sentence-structured approach. The initial phase of outlining is the one time in writing when the rules of proper grammar can be pushed aside. The objective of the initial outline is to identify the important points that you wish to communicate, not to produce a grammatically correct document. The first outline might be a bulleted list that identifies:

- The problem
- The specific goal that underlies the work
- The approach used in the work
- One or two important results
- A major conclusion

Immediately after completing the first outline, you can expand it into a more detailed second outline, which will provide slightly more detail and begin to provide some structure for a rough draft. As with the first outline, you should use brief phrases, not complete sentences.

Subsequent outlines will provide progressively more detail and greater structure. The expansion of the outline of the introduction can give greater detail about the problem and greater specificity of the objectives. Expansion of the literature review section of the outline can cite specific authors, their methodologies, and the input requirements for their models. The outline for other sections of the report may be separated into parts with headings and subheadings used.

Outlines increase the efficiency of report writing and also improve the quality of the final report. Outlining helps the writer focus on the important points, as well as ordering the topics in a logical sequence. An outlined report will produce a report that the reader will find easier to follow.

A.1.2 ROUGH DRAFTS

Just as getting started with the first outline may have caused some apprehension, getting started with the first rough draft may elicit similar feelings. Starting with the introduction is not recommended. It is probably best to start with the section of the report that appears to be the easiest to write. Very often, this is the literature review or data analysis section. Writing the least likable parts can be put off until drafts of the easier parts have been completed. Starting with the easier parts will generate positive feelings about completing an entire rough draft.

In the initial rough draft, do not worry about smooth transitions between paragraphs and sections. The transitional sentences can be inserted in subsequent drafts. The following are some additional tips for making a rough draft:

1. Do not worry about specific wording, spelling, or punctuation—these can be cleaned up in subsequent drafts.
2. Write quickly with the purpose of getting ideas onto the paper—avoid stopping; otherwise, writing momentum will be lost. In the rough drafts, do not stop to critique your ideas, as self-criticism will reduce your motivation.
3. If motivation to complete a section wanes, start work on another section, so that writing momentum is not lost.
4. Do not be critical of your first drafts. Critical assessments will divert your attention away from the task of getting words down on paper.
5. Do not stop to draft figures or tables when completing the rough draft—just indicate that a table or figure is needed at a certain point and make notes on the important elements to be included in the table or figure.
6. Do not be tied to your outline. As you write, expect that new ideas will surface; changes to the outline are expected and can lead to improvements in the paper.

A.1.3 REVISION

Once the first rough draft is completed, put it aside for a few days. The hiatus will help you to take a more independent view of your writing. In a way, it will be similar to having a friend review your work. You should not expect to go from a rough draft to a final draft in one step, and some parts of a draft may require more revision than others. Good writers recognize the need for multiple drafts. Writers often approach each revision with a different purpose. One revision might focus on the transition sentences. Another revision might focus on ensuring that each paragraph includes a good topic sentence, while another revision could be devoted to making sure that the implications of the results are addressed. At some point, a revision should ensure that the sentences are clearly written and that all misspelled words are corrected. The following list includes items that should be considered in revising a report:

1. Objectives are clearly stated in the introduction
2. Headings throughout the report are numerous and descriptive
3. All technical terms are defined
4. All notation is defined
5. Words are chosen properly
6. Gender-neutral wording is used
7. Sentences are structured properly
8. Proper transitions are made between ideas
9. Paragraphs are structured properly
10. Each paragraph includes a topic sentence, with each sentence in the paragraph related to the topic sentence
11. Analyses are unbiased
12. All figures and tables are included
13. Each figure and table can be independently understood
14. Each figure and table has a descriptive title and is numbered appropriately
15. All ideas taken from other sources are properly referenced

When you are satisfied with the report, you should have someone else critique it. It is important to enlist the help of someone who will be very critical of your writing and ideas. It does little good if the reviewer who you select just returns it to you without criticism. Too often, the writer forgets that he or she is totally familiar with the work; thus, details that are necessary for the intended reader may have been omitted. Just as it is helpful to have a friend listen to a rehearsal of a speech, a friend's review of a written report can lead to many positive changes.

A.1.4   Paragraph Structure

Paragraphs should be more than a group of sentences. Unfortunately, writers put together five to eight well-written sentences but one poor paragraph. The difficulty is that the grammatically correct sentences are only peripherally related and not even well ordered. The sentences seem to be slightly connected, but not such that they belong in the same paragraph.

The problem results from a lack of outlining and failure to recognize the importance of a topic sentence. A good outline would suggest a topic sentence for each paragraph. A second part of the problem is improper proofreading. The sentences in a paragraph are reviewed to identify misspelling and poor grammar, but not for cohesion among the sentences. One revision of a draft should be devoted to checking each paragraph, to ensure that all sentences in the paragraph are relevant to the idea behind the topic sentence. This one step can lead to a vastly more readable report or paper.

A.1.5 ABSTRACT

An abstract is a vital part of any technical report; in fact, it may be the most important part. In professional practice, the abstract is sometimes called an executive summary. Readers use the abstract to evaluate the relevance of the document. It is often the first part of the report read by anyone interested in the subject. It is, in effect, an advertisement for the report. A well-written and complete abstract can sell the importance of the information reported in the document to the potential reader. Abstracts and executive summaries are also used as a substitute for reading the entire formal report.

A well-written abstract will identify the following: (1) the underlying problem (i.e., why such work was needed), (2) the objectives of the work (i.e., what the work intended to accomplish), (3) the scope (i.e., the parameters of the study), (4) a brief statement of the results (i.e., how the work fulfilled a need), and (5) a discussion of the implications of the results (i.e., how the work advances the state of the art or the value of the work to society). The abstract must be more than just a summary of the paper. A good abstract will entice the reader to read and subsequently use the material in the report. A poorly written abstract will suggest that the report is also poorly written, thus discouraging the reader from reading the report. References should not be cited in an abstract because an abstract is meant to stand alone. Acronyms should be avoided.

Executive summaries (ESs) are generally much longer than abstracts. Whereas an abstract might be 100 to 150 words, an ES might be 1,000 words. The important point to remember about an executive summary is that it is like an advertisement for a consumer product. The latter has the objective to get the reader of an advertisement to buy the product. The ES has a very similar objective, namely, for the ES reader to "buy" the report, i.e., to read the entire report. The reader will be more intrigued if the ES concentrates on the knowledge that he or she will gain from reading the entire report. Therefore, a long discussion of the problem is not necessary; instead, most of the length of the ES should be an informative summary of the conclusions and implications of the study. Too frequently, the ES concentrates on the statement of the problem rather than the important conclusions. Such an ES will not entice the ES reader to read the entire report, as he or she is already familiar with the problem.

The following summarizes the content of an informative ES or abstract and the appropriate percentage of its length that should be devoted to the different parts:

- Statement of the problem (10%)
- Statement of the goal and objectives (10%)
- The methodology used, i.e., the analyses (20%)
- Results (30%)
- Conclusions and implications of the work (30%)

A.1.6 HEADINGS

Headings are a very important characteristic of a report. Can you imagine trying to read this book if the headings and subheadings were omitted? Headings help the writer organize the report and are essential for the reader to follow the content of the written document. Headings must be descriptive, yet concise. The following examples show inadequate headings and descriptive alternatives:

Not Descriptive	Descriptive
Model Development	Formulation of Transportation Planning Model
Data Analysis	Statistical Analysis of Maryland Water Quality Data
Results	Design Curves for Estimating Bridge Scour
Model Application	Application of CREAMS Model to Agricultural Areas
Recommendations	Policy and Design Recommendations
Discussion	Implications of the Model Study to Engineering Practice

In each of these comparisons, the descriptive heading is longer than its counterpart. While excessively long headings should be avoided, it is reasonable to have headings that are fifty to sixty characters in length. Each heading should tell the reader about the general content of the paragraphs that follow the heading.

The format of the headings is also very important. Since the headings and subheadings guide the reader through the report, the format should be clear and consistent. Several format structures are possible. Reports often use a numbering system, such as:

1.1
   1.1.1
   1.1.2

1.2
2.1
2.2
   2.2.1
      2.2.1.1
      2.2.1.2
   2.2.2
2.3

The two-number entries (e.g., 1.1 and 2.3) are the primary sections of the chapter, which is indicated by the first number; for example, 5.3 would indicate the third major section of Chapter 5. Subsections of major chapter sections are indicated with three or four numbers. Figure A.1 shows another option for structuring headings.

A.1.7 THE REPORT INTRODUCTION

The introduction should include a concise statement that defines the problem, a brief history leading to the problem, and the purpose of the work. Specifically, the overall goal of the report and several objectives should indicate the purpose of the report. The introduction should not be a review or summary of the report, and it must not

One possible arrangement of titles and subdivisions within a chapter and their spacing is illustrated as follows:

FIRST-ORDER HEADING

If only one rank of heading is used within a chapter, subdivisions should be indicated by a statement in all capitals and often in bold script, with two spaces between it and the last line of text above ("two spaces between" means that the typist triple spaces) and one space between it and the text following (i.e., the typist double spaces).

Second-Order Headings

If two ranks of headings are used, subdivisions within the main rank are indicated by an underlined heading with one space above and below, and in initial capitals. The headings begin at the left margin. Underlining the second-order heading is optional.

Third-Order Headings. If three ranks of headings are used, a third-order heading is indicated by a heading indented five spaces (the number of spaces indented for a regular paragraph), underlined, with initial capitals, followed by a period, and with the paragraph beginning on the same line and two spaces after the period.

FIGURE A.1   Possible heading format for formal reports.

contain conclusions of the report. Instead, the introduction should set the stage, define limits, and describe why the work is needed. The underlying purpose of an introduction is to entice the reader to continue reading, not give results that would discourage the reader from continuing to read.

A.1.8   CONCLUSIONS

This section should summarize the results of the work presented in the report. The results should be based on factual findings. Each separate conclusion should be discussed in a logically sequential order. Because conclusions are often lifted out of the context of the report and quoted without the explanatory material, the writer should take care to compose each conclusion to ensure that it does not imply a broader scope than is intended, and that it includes the necessary qualifications. *The tendency to present the conclusions in outline or bullet list form should be avoided* because it is important to state the implications of the findings. Thus, one-sentence statements summarizing the important findings may be inadequate to portray the scope of the investigation.

A.1.9 APPENDICES

The purpose of an appendix is to present those details of data or methodology that will verify the summary statements reported in text. The material is not placed in the main body of the report because it would obscure the development of the presentation. The appendix or appendices should contain the bulk of the data or findings as embodied in the tables, diagrams, sketches, curves, and photographs. They should contain such items as sample computations and derivations, computer programs and output, and material that is too voluminous for inclusion in the main report. A "Glossary of Abbreviations" may be included as an appendix. Also, an appendix entitled "Notation" may be used to list the definitions of all mathematical symbols used in the report. Each appendix should be indicated by a letter (e.g., Appendix A), include a title, and have a cover page.

A.1.10 GRAPHICAL COMMUNICATION

As the saying goes, "a picture is worth a thousand words." Graphs are an important part of technical communication. In fact, technical prose sans accompanying graphs can be ineffective in communicating important results. Numerous methods are available, including histograms, box plots, and x–y plots, with the first two used with a single random variable and the latter with two random variables. Other types of graphical analyses include line graphs, bar graphs, and polygons.

Histograms: A histogram graphs the frequency or relative frequency of occurrence of sample data. The relative frequency is obtained by dividing the actual frequency by the total frequency, which yields values from 0 to 1, which make comparisons easy. The following are guidelines for constructing an effective histogram:

1. Set the minimum value of the random variable X, as X_{min}, as either (a) the smallest sample value or (b) a physically limiting value, such as zero.
2. Set the maximum value, say X_{max}, as either (a) the largest sample value or (b) a physically limiting value.
3. Select the number of intervals or cells, n_c; an empirical estimate is $n_c \sim 1 + 3\log_{10} n$, where n is the sample size.
4. Compute the approximate class width ($\hat{w}$) as $\hat{w} = (X_{max} - X_{min})/n_c$.
5. Set the actual class width (w) by rounding $\hat{w}$ to a convenient value.
6. Identify the cell boundaries (b_i, $i = 1, 2, \ldots, n_c$) by $b_i = X_{min} + i_w$. Note that the first of the n_c cells ranges from X_{min} to $X_{min} + w$.
7. For each cell count the number of sample values within its bounds.

It is necessary to use judgment in setting the number of cells and the bounds. It is often a good idea to graph a couple of histograms using different cell bounds in order to see if they lead to the same interpretation.

Illustrations: Illustrations, except for tabular data, should be labeled as figures. Many of the rules for tables are applicable to illustrations. The categories of illustration can include figures, nomographs, photographs, or descriptive summaries not in tabular form. Figures should be designated by an Arabic numeral and numbered consecutively throughout the report. For reports that are subdivided by sections or chapters, the section or chapter number may be included as part of the figure number. On the figure itself and when referring to the figure, the word *figure* should have the first letter capitalized.

In addition to being numbered, each figure requires a descriptive title, placed at the bottom of the figure using first-letter capitals. Like tables, figure titles can be used to describe the content of the figure, define notation, and specify units of variables, if the units are not specified on the axes.

The axes of a figure should also be labeled, including the units of the variables. For figures with multiple lines, a description of the individual lines should be placed next to the line, unless it is part of the title or included in a box within the figure itself. Examples of properly structured figures can be found in any textbook.

Except in special cases, illustrations should be black and white. Colored lines should not be used because they are indistinguishable on photocopies. For multiple-line illustrations, various combinations of broken lines can be used to distinguish them. A few such forms are (———), (– – – –), (------), (–•–•), (–••–).

Pie charts: When data are expressed as percentages, proportions, or fractions of a whole, pie charts can be used to enhance the material. A round circle is used to represent 100% and the "pie" is sectioned according to the percentages. The pie chart has the advantage that the size of the pie slice supports the numerical values.

Box-and-whisker plots: Box-and-whisker plots are a graphical method for showing the distribution of sampled data, including the central tendency (mean and median), dispersion (10th, 25th, 75th, and 90th percentiles), and extremes (minimum and maximum). Additionally, they can be used to show the bias about a standard value and the relative sample size, if the figure includes multiple plots for comparison.

To construct a box-and-whisker plot, compute the following characteristics of a data set:

1. The mean and median of the sample
2. The minimum and maximum of the sample
3. The values that 90, 75, 25, and 10% of the sample are less than or equal to

The plot consists of a box, the upper and lower boundaries that define the 75th and 25th percentiles, and upper and lower whiskers, which extend from the ends of the box to the extremes. At the 90th and 10th percentiles, bars, which are one-half of the width of the box, are placed perpendicular to the whiskers. The mean and the median are indicated by solid and dashed lineds that are the full width of the box.

If a figure includes more than one box-and-whisker plot and the samples from which each plot is derived are of different sizes, then the width of the box can be used to indicate the sample size, with the width of the box increasing as the sample size increases.

A.2   GUIDELINES FOR ORAL PRESENTATIONS

The oral communication process can be separated into four steps: (1) formulating the presentation, (2) compiling the material for the presentation, (3) rehearsing, and (4) making the presentation. It is important to recognize that a poor presentation almost always results from failure during the first three steps, not the fourth step. If sufficient attention is given to the preparatory steps, then the actual presentation will most likely be successful. Proper attention to the first three steps can also help reduced nervousness, which is usually the number one concern of the novice.

A.2.1   FORMULATING THE PRESENTATION

The best way to formulate a presentation is to answer the question, "What major point(s) should be made?" By focusing on the major conclusions of the presentation, one can then prepare to educate the audience.

- *Know your audience.* A speech to be presented to a homogeneous audience will be different from one prepared for a heterogeneous audience.
- *Educate your audience.* In preparing the speech, identify the educational objectives of your presentation. What new knowledge do you want the audience to have after listening to your speech?
- *Entertain your audience.* The audience is more likely to grasp the educational points of your presentation if it is presented in an entertaining fashion. Effective visuals provide the opportunity for making a presentation more entertaining, but make sure that the approach to entertainment will be well received by the audience and not viewed as sophomoric.
- *Persuade your audience.* Your presentation should include strong supporting material in order to persuade the audience to believe in your conclusions.
- *Be mindful of time constraints.* When formulating a presentation, know the time allotted to the speech and plan only to develop material that can be effectively presented in the allotted time.

A.2.2   DEVELOPING THE PRESENTATION

A very efficient way of developing a presentation is to use the progressive outline approach. With this method, a very simple outline of four to six lines is made to address the following questions:

1. Why was the work done? (State problem and goal.)
2. How was the work done? (State solution method.)
3. What findings resulted from the work? (State one or two major conclusions).
4. What do the results imply? (State the implications of the work.)

Guidelines related to these questions are as follows:

- *Content of subsequent outlines.* With each outline, add more content that relates to the educational objectives identified in the speech formulation phase.

- *Begin organizing visuals.* Visuals add variety to a presentation and can serve as the medium for entertainment and education. Visuals outline the presentation for the audience and can serve as cues for the speaker.
- *Give special attention to the introduction.* Nervousness is most severe at the beginning of a speech, so a well-developed introduction can increase the speaker's confidence. Also, a poor introduction will cause the audience to reduce their attentiveness.
- *Focus on the conclusions.* The conclusions to a speech are important because the end of the speech is when the speaker identifies the major points. The conclusions are the points that the audience should learn from the speech.
- *Capture the audience's attention.* Make sure the major points are covered without trying to do too much. Presentations that are crowded by too many details are often ineffective.

A.2.3   REHEARSING THE PRESENTATION

Rehearsing is important to reduce nervousness, to ensure that the time constraint will be met, and to become sufficiently familiar with the material that notes will not be needed. A few guidelines related to rehearsal are as follows:

- *Location, location, location.* Try to practice in the exact place where the speech will be given. Familiarity with the surroundings helps put a speaker at ease.
- *Practice the opening statement.* Nervousness is greatest at the start of a speech, so be extra familiar with the opening remarks to ensure that these capture the attention of the audience.
- *Have friends critique the presentation.* Rehearsing with an audience rather than by yourself will make the rehearsal like the actual presentation. The friends should also be willing to give serious criticism so changes can be made before the presentation.
- *Do not rehearse in silence.* When rehearsing, speak aloud. If you just rehearse by mouthing the words, you will not be able to judge the time because you speak slower when speaking aloud than when rehearsing in silence.

A.2.4   MAKING THE PRESENTATION

If you were successful in the first three phases of the oral communication process, then chances are that the actual presentation will be successful. A few guidelines relative to making the presentation are:

- *Nervousness is good.* Some nervousness is to be expected and can be beneficial if it makes you concentrate more on your presentation and less on the audience.
- *Make eye contact.* Making eye contact with those in the audience will help keep them engrossed in your presentation; however, you do not want to think about the person with whom you make eye contact.

- *Avoid filler words.* Words like *uhm*, *uh*, or *you know* are called filler words, as they are spoken, often unknowingly to the speaker, to fill the time gap between ideas. Short gaps of silence are not necessarily bad, as they give those in the audience time to think about what you have said. However, continued use of filler words can be distracting to the audience.
- *Be careful of bad body language.* Your hands can be distracting to the audience if you motion too much or not enough. Also, do not fold your arms in front of you, as this indicates that you are closed to the audience. Keep your hands out of your pockets. Good body language can help your presentation succeed.

A.2.5   RESPONDING TO QUESTIONS

The question-and-answer period is very important, and questions should be encouraged. Questions show that you have sparked their interest. When you are asked a question, do not be afraid to pause for a moment to formulate a response. Try to keep your response short and to the point. If you are asked a question and have no idea how to respond, ask the person to rephrase or clarify the question. If you understand the question but do not know the answer, be honest and state that you do not know. If a person seems combative with the question, do not try to match his or her attitude; it is better to respond in a very neutral tone and move on to another question.

Appendix B: Creativity and Innovation

APPENDIX OBJECTIVES
- Define creativity and innovation
- Identify the importance of creativity and innovation
- Present creativity stimulators
- Discuss creativity inhibitors

B.1 DEFINITIONS

The word *create* has the meaning to originate, to bring into being, or to produce. It is generally assumed that creating means that the product was the first of something. For example, Rutan created the first airplane that was flown solo around the globe without refueling. The word *creativity* is applied to individuals who have the ability to create. They have the imaginative powers, i.e., mental processes, and the attitude to recognize important problems and to develop new solutions to solve the problems.

The word *innovation* means to alter, change, adapt, modernize, or add to. Some liken innovation to creativity, but the two differ in that creativity is generally applied to the development of something new, while innovation refers to making a change to an existing product. The line between the two is not always obvious. For example, the first handheld vacuum cleaner was a creative product, and therefore some may consider it only an innovation of the regular upright vacuum cleaner. However, adding the rotating brush to the handheld vacuum is clearly just an innovation of the brushless handheld vacuum.

Both creativity and innovation are important to the civil engineering profession. Much of civil engineering represents creative development of solutions to engineering problems. The development of new materials, new structural design methods, and computer algorithms to solve problems in a unique way would be considered creative. Modifying an existing construction practice or adapting the principle of in-stream storm water detention to out-of-stream storage would represent innovation. Both creativity and innovation are necessary for the civil engineering profession to meet its responsibilities to serve society and ensure public health, safety, and welfare.

B.2 THE CREATIVE PROCESS

The creative process is a way of solving a problem in an original way. Using this process generally leads to a useful solution or product. The scientific method is usually viewed as including the following four steps: observation, hypothesis, experimentation, and induction. The creative problem-solving process can be viewed in four corresponding steps:

1. *Recognition*: A period of recognizing the problem and questioning the reasons for the problem.
2. *Selection*: The phase of identifying the most appropriate creativity stimulator to use to generate potential solutions to the problem.
3. *Ideation*: The phase where one's imaginative powers are used to generate ideas that may stimulate a solution.
4. *Evaluation*: The phase of critically evaluating the ideas of the third phase and developing a realistic solution to a problem.

The literature on creative thinking generally concentrates on phase 3, the idea generation phase. However, recognizing the real problem and turning ideas into reasonable solutions are just as critical to problem solving. Asking questions is a central element in all phases of the creative process.

B.3   MYTHS ABOUT CREATIVITY

The topic of creative problem solving has not been without its detractors. These individuals have voiced opinions that have led to myths, two of which will be addressed.

The first myth says that generating wild and crazy ideas is fun, but it is not productive. Yes, methods of creativity, such as brainstorming, often involve generating ideas that lack an obvious connection to the problem at hand. The idea of playing sexy music to make a banana peel its skin may not seem related to the problem of removing an undesirable environmental slime from vegetation, but the idea may encourage a different viewpoint on the problem. The sexy music focus may encourage more general questions and prompt the problem solvers to take a broader perspective on potential solutions. Eventually, the idea of using sound waves to remove the slime may result from the idea of the music. This is generally the benefit of applying methods of creativity, i.e., creativity stimulators such as brainstorming.

A second myth, regarding the topic of creative thinking, is that you either have creative ability or you don't; it is not something that you can learn. This is not true any more than it is true that you were or were not born with the ability to play soccer or the piano or do calculus. Certainly innate characteristics influence one's creative intelligence quotient (CIQ), but that does not imply that a person cannot drastically increase one's CIQ. Just as gaining knowledge and experience at playing piano can increase one's ability at that activity, being more knowledgeable about creativity stimulators and believing that they can help problem solving will enable one to use creativity stimulators more effectively.

B.4   THE IMPORTANCE OF CREATIVITY

Research, even if it is aimed at making minor improvements to an existing idea, is never problem-free. Just getting an idea for research can be a struggle. All other aspects of the research process are subject to problems. For example, extreme events in measured laboratory data are not uncommon. The inability to control an environmental factor in field research can introduce considerable random scatter into data. Where it is necessary to build apparatus for conducting an experiment, it is often problematic in getting the apparatus to meet all of the necessary research needs. Problems such as these, and many others, require good problem-solving skills. This is where creative thinking, or creative problem solving, can be of value.

Your creative problem-solving skills are valuable tools in many aspects of engineering, such as

- Finding and developing a topic to research
- Developing an efficient experimental design that will meet all research objectives
- Solving problems that arise when conducting the experiments
- Identifying general implications of the research
- Preparing effective oral presentations of the research
- Interpreting research results that do not match initial expectations

If a research topic is worthy of investigation, then problems will arise. Knowing creativity stimulators, such as brainstorming, and creativity inhibitors can increase both the efficiency of problem solving and the likelihood of success.

B.5   CREATIVITY STIMULATORS

At some point in time, each of us has probably brainstormed to solve a problem, likely without realizing that we were doing it. Brainstorming is one method of generating ideas for the purpose of finding a useful solution to a problem. It is one of a number of methods called *creativity stimulators*. Three creativity stimulators are presented: brainstorming, brainwriting, and synectics. The latter two are variations of the general methods of brainstorming.

B.5.1   BRAINSTORMING

The basic idea behind brainstorming is the generation of ideas. Generally, a group selects a facilitator whose job it is to (1) encourage idea generation; (2) record the ideas, preferably where everyone in the group can see the ideas; and (3) contribute ideas, especially when the group is struggling to produce new ideas. A white board, chalkboard, or overhead projector is the usual alternative for recording the ideas.

Effective brainstorming sessions depend on an active, encouraging facilitator and a group that adheres to the following rules for brainstorming:

1. The more ideas produced, the better
2. The wilder the idea, the better
3. Combination and improvement of ideas are encouraged
4. Criticism of ideas is unacceptable

The brainstorming session begins with the facilitator stating the problem in general terms. A general statement is preferable to a specific statement, as it encourages a broader array of ideas. The group should be in a room conducive to idea generation and where everyone can easily see the ideas being recorded by the facilitator. It is important to exclude those individuals who do not believe that brainstorming will lead to an acceptable solution. They act as creativity suppressors and significantly decrease the effectiveness of the group.

Criticism of ideas, regardless of how wild and crazy an idea is, must be suppressed, as it acts to inhibit the generation of more ideas and is deflating to those with positive attitudes. One job of the facilitator is to judiciously discourage those who criticize ideas from inhibiting the momentum of the group.

Once a reasonable list has been compiled by the group, it is the responsibility of the facilitator to begin the evaluation phase of the brainstorming process. Each idea is analyzed by relating it to the problem for the purpose of seeing if it can lead to a reasonable solution. To encourage the group toward actively participating in this phase, the facilitator can ask the group to select the "wildest and craziest" item on the list. By turning this idea into a possible solution, the group will more actively analyze the other items on the brainstorming list.

Table B.1 includes the partial results of two brief brainstorming sessions, where the length of each list was limited to ten items. Some ideas build on a previous idea (e.g., the idea of riding a whale is similar to that of riding of a dolphin). Many of the ideas are far-fetched (e.g., beam yourself across or play sexy music to make the banana strip), while others are very practical (e.g., cut it off with a knife). Both crazy and practical ideas should be encouraged.

Useful solutions can come out of the wild and crazy ideas. For example, the use of sound waves might be a solution that follows from the "play sexy music and strip" item. Strings laced with an acid placed on the seams of the peel may prove to be a useful solution.

TABLE B.1
Examples of Brief Brainstorming Sessions

Ways of Crossing a River

- Shoot yourself out of a cannon
- Tie yourself to a dolphin
- Ride a whale across
- Use a pogo stick that freezes the water below it
- Swim across
- Build a bridge
- Tunnel under the river
- Catapult yourself across
- Beam yourself across (as in Star Trek)
- Suddenly stop the Earth from turning so your momentum will carry you across

Ways of Peeling a Banana

- Give it to a monkey after taping its mouth shut
- Attach explosives to the seams
- Genetically create a banana without a peel
- Play sexy music and make it strip
- Design a machine to do it
- Use a special acid that would destroy the peel
- Burn it off with fire
- Cut a hole in the top and squeeze it out
- Vaporize the peel with a laser
- Cut it off with a knife

B.5.2 BRAINWRITING

While problem solving most often is a team activity, individuals must often solve their own problems. This is especially true for problems that involve multiple disciplines. The individual must solve problems in his or her own specialty. Creativity can just as much be an individual's activity as it can be a group effort.

Brainwriting is nothing more than one-person brainstorming. The individual identifies the central problem, and then generates ideas. Instead of having a facilitator as in group brainstorming, the individual acts as both the brainstormer and the facilitator. The ideas should be recorded while they are generated. While recording ideas, it is very important to abbreviate the idea. The recorded idea should only be long enough to ensure that the meaning behind the phrase is not lost. It is also important to keep from being critical of ideas as they are generated, which can be a tendency when brainwriting. A large quantity of ideas is still wanted when brainwriting.

B.5.3 SYNECTICS

Synectics is a form of brainstorming where the group does not know the real problem. Instead, they are presented with a generalized problem that has some relationship with the specific problem. The group then brainstorms on the generalized problem. When a reasonable list of ideas has been generated, the next step is to develop an association between each item on the brainstorming list and the specific problem. These associations are then used to identify solutions to the problems.

The following are some examples of specific problems and generalized problems:

Specific Problem	Generalized Problem
Improve lawn mower cutting	Improve knife to cut vegetables
Improve street lighting	Build a better reading light
Improve street lighting	Get more sunlight into homes
Improve sound quality of stereos	Improve mail delivery

Note that more than one generalized problem can be identified for any specific problem, as with the street lighting example.

Assume that improving the sound quality of stereos is the specific problem and that the group is asked to brainstorm ways of improving mail delivery. The following shows a few ideas generated and their association with the sound quality problem:

Brainstorm Idea

- Reduce walking distance
- Optimize shoe traction on type of surface on which the mail carrier walks
- Better control of dogs
- Larger door slots

- Minimize length of connecting wires
- Reduce noise in the system

- Recommend ways to improve sound in the room
- More speakers, each with smaller frequency range

Obviously, "better control of dogs" is the wildest idea generated. However, it can lead to a solution that may be more effective than making modifications to the stereo system, as it may provide high-quality sound. The control of dogs can be considered an environmental factor, as the dog is not part of the normal delivery process. Adjusting the layout of the room where the stereo is played may be more effective than modifying an already high-quality stereo system.

Synectics has the advantage that the group does not know the real problem, so they are more likely to be open-minded and generate a wider array of responses. The group will be less likely to prejudge ideas or try to find very practical solutions at the expense of creative solutions that might prove to be more effective.

B.6 CREATIVITY INHIBITORS

To be successful in the use of creative problem-solving methods often requires personal change. Many internal fears and attitudes prevent an individual from reaching his or her full creative potential. Overcoming these fears and negative attitudes is often the first step to becoming a more creative problem solver.

Table B.2 lists a few creativity inhibitors, which are separated into those that are personal fears or attitudes, and external inhibitors that reflect attitudes of those with whom we must interact. A fear of criticism by others can discourage a person from applying his or her creative powers. A fear of making mistakes, or even of failure, can make a person disinclined to risk applying his or her creative ability. Success often brings notoriety, which some people shun; this discourages them from implementing creative solutions to problems.

TABLE B.2
Creativity Inhibitors

Internal

Fear of:

- Criticism
- Making mistakes
- Failure
- Success

Attitudes

- Lack of confidence
- Unquestioning
- Pessimism
- Self-critical
- Impatience

External

- Ridicule by others
- Unwillingness of others toward change
- Superiors who delay implementing new ideas

The lack of self-confidence may be the most common and most detrimental inhibitor. A belief in one's own ideas and problem-solving ability is a fundamental necessity to research. Someone who lacks confidence must overcome this attitude by acknowledging past successes and by taking the "if others can do it, I can do it" attitude. Even small successes in the past should be viewed as examples that success is possible. Developing self-confidence takes time, but the active use of your creative powers will enhance your self-confidence. The two attitudes are synergistic.

Problem solving in research involves questioning. For example, "Why won't this work?" and "What is the real problem?" are two questions fundamental to research. If the real problem is not understood, then the real solution cannot be identified. Identifying the real problem will be hastened by asking questions, such as "Why am I getting a positive relationship when I expected a negative relationship between the two variables?" A person is more likely to seek answers if the problem is posed as a question. Very often the first question is not the important one, but in trying to answer the first question, subsequent questions and answers will lead to the correct question, and then the correct solution to the problem. For this to happen, a questioning attitude is important.

A pessimistic attitude can be a real inhibitor to the successful use of creative thinking techniques. Pessimism is to creative thinking as a lack of confidence is to the individual. A user of brainstorming must have the confidence that it can help solve the problem. The belief that the problem cannot be solved will discourage the use of creative problem-solving techniques and the asking of questions. Pessimism is correlated with a self-critical attitude. One who is pessimistic will generally be critical of ideas, thus taking the attitude, "It won't work!"

Solving real problems takes time, as well as creativity. A person who wants or expects a quick solution will generally not find the best solution. Therefore, impatience is a creativity inhibitor. Yes, impatience will promote efficiency if it promotes concentration and diligence, but if it hinders a person from thoroughness, then the downside can outweigh the upside.

In addition to the personal fears and attitudes, external or environmental factors can limit creativity. Critical comments from superiors on your ideas can decrease your self-confidence and willingness to recommend any solution other than the obvious ones. Criticism can be a positive if it can lead to new knowledge or a clarification of the problem. Criticism is a negative when the reason for the rejection of an idea is not also communicated. Just saying that it is a lousy idea will actually move the group further from the solution because it discourages active participation.

B.7 DISCUSSION QUESTIONS

1. What are the characteristics of a creative person? Explain each.
2. Obtain a definition of *creativity* from a dictionary. Discuss how it applies to research.
3. Discuss the use of the four steps of the creative problem-solving process in research.

4. How can wild and crazy ideas be useful in research where practical solutions are the goal?
5. How does an optimistic attitude increase the likelihood of success in research?
6. Brainwrite for ten minutes on the topic of ways of opening a tin can when a can opener is not available.
7. Brainwrite for ten minutes on the topic of ways of increasing home security.
8. Brainwrite five ways of improving mail delivery. Then use these five responses to improve the efficiency of a robot.

Index

9.4.5 Associate with Professional Societies

Working with those from a technical society, such as ASCE, to meet with legislators is another option. This has the obvious advantage that the legislator is able to see that the "special interest" group has the public welfare at heart and the technical knowledge and experience that justifies its position on the legislation. This has shown to be an option that can influence public policy at all levels of government.

9.5 ANALYSIS OF A SAMPLE POLICY STATEMENT

Consider the following sample policy statement on storm water management.

POLICY STATEMENT

It is the policy of the administration (1) to minimize loss of life and property from flood damage by promoting state and local programs that prevent the development of new damageable property, (2) to assist in the development and construction of sound, cost-effective flood control structures, (3) to implement a storm water management program that will effectively prevent an increase in the magnitude and frequency of flood flows, thus preventing an increase in flood hazard, (4) to maintain the integrity of the natural stream channel geometry, and (5) to encourage the design and implementation of storm water management systems that minimize the entrainment of pollutants and provide a reasonable degree of control of storm water before runoff reaches the stream system.

DISCUSSION

The optimum design of storm water runoff collection, storage, and conveyance systems should simulate as closely as possible the features and functions of the natural drainage system that are largely capital-, energy-, and maintenance-cost-free. The system selected should balance capital costs, operation and maintenance costs, public convenience, risk of significant water-related damage, pollution prevention, fish and wildlife habitat preservation, environmental protection or enhancement, and other community objectives. The choice of a solution to any storm water management or flood-related problem depends upon the situation at the time the problem is being addressed. In the past, solutions to storm-runoff-related problems have had undesirable and lasting environmental, economic, and social effects. The selection of certain methods and structures can destroy environmental features that are of value to the community. The construction cost can be high, and the long-term maintenance costs may continue to burden society.

This policy statement can be evaluated from a number of perspectives. Values and technical issues within this policy will be presented; however, criteria such as costs, aesthetics, pollution abatement, and environmental health can also be used to evaluate a policy.

9.5.1 PUBLIC VALUES

A number of human values are evident in the above policy statement:

- To minimize the loss of life and property (public safety)
- To design and construct sound flood control structures (safety)
- To prevent an increase in flood hazard (public safety)
- To maintain the integrity of the natural stream (sustainability)
- To minimize the entrainment of pollutants (public health, sustainability)
- To provide public convenience (efficiency)
- To provide environmental protection (sustainability)
- To minimize long-term maintenance (public welfare)

This part of the overall policy is very value laden, which is just one reason to have a deep appreciation for human values. The technical part of the policy statement, which is not shown, provides the technical details necessary to meet the value responsibilities stated throughout the policy.

9.5.2 TECHNICAL DIRECTIVES

In addition to the values, policies are heavily laden with technical issues:

- To prevent the development of new damageable property (land use planning)
- To design and construct sound, cost-effective flood control structures (efficiency, engineering economics)
- To prevent an increase in the magnitude and frequency of flood flows (statistical hydrology)
- To provide a reasonable degree of control of storm water before runoff reaches the stream system (environmental planning, best management practice)

A full policy statement would, of course, include many details about the way that the flood control structures would be designed, constructed, maintained, and financed. The important point is that a public policy addresses much more than technical issues, which are typically the domain of the engineer. To fully appreciate the intent of a policy, the engineer must recognize human values and use them to guide the development of the technical details.

9.6 ETHICAL ISSUES RELEVANT TO PUBLIC POLICY

Ethics are usually associated with topics like conflict of interest, kickbacks, plagiarism, and falsifying documents. These four ethical problems are often associated with both legal and ethical issues, so the relationship of these issues and engineering practice is obvious. This is not the case with issues that are relevant to public policy.

Even though legal issues may not have a dominant role in the engineer's daily workload, engineers regularly deal with public policies and the ethical issues inherent to them. Thus, ethical issues are very relevant to engineering work.

When an engineer has technical expertise and experience relevant to a proposed public policy or an existing policy, he or she also has value responsibilities to the public. A failure to participate may suggest an insensitivity to the value responsibilities addressed in professional codes of ethics. These values may include sustainability and public health, safety, and welfare.

On the issue of sustainability, each individual will need to recognize its importance, including the long-term effects on future generations. Sustainability will not succeed unless each person does his or her share. However, individuals cannot make it succeed without the public policies that give direction. Engineers and the engineering profession have important ethical responsibilities related to sustainability. More than any other profession, civil engineering has significant technical expertise relevant to sustainability, such as energy production and distribution, the understanding of environmental processes, recycling and waste management, and natural resource processing. Issues such as these will require public policies that address the technical issues, and civil engineers will be central to the solutions because of their technical expertise.

A researcher who does not communicate the results of his or her research to the public or the engineering community is failing to transmit knowledge to those who may need it. This is often viewed as being selfish, but it is not illegal. However, ethicality is not a yes or no issue. Like the researcher, an engineer who has knowledge that could help society through public policy formation could be failing to communicate knowledge that would benefit society. Is it any less ethically questionable to fail to communicate knowledge to help public policy than to do the same for the research? In both cases, knowledge is a key value, and its dissemination for public policy formation or technology expansion is an ethical issue.

What should engineers do if they believe that a public policy, such as one that does not provide adequate support for public infrastructure, is not in the best interest of society? Do they have an ethical obligation to react to the situation, and if so, what action should they take? All codes of ethics point to the engineer's responsibility to public health, safety, and welfare. A lack of state-of-the-art infrastructure certainly affects public safety. This suggests that the engineers must take action to get poor public policies corrected; otherwise, they would be failing in their ethical responsibilities to the public. The actions can include those discussed previously. Letters to the editor, meetings with congressional delegates or state legislators, and community group meetings are avenues for change. In most cases, groups of engineers must be coordinated to achieve significant change.

9.7 DISCUSSION QUESTIONS

1. Obtain a copy of the Clean Water Act and summarize the factual content of the act.
2. Obtain a copy of a state policy related to an environmental issue (e.g., controlling discharges into the Chesapeake Bay) and summarize the factual content of the policy.

3. Identify a recently passed bill that is related to engineering for your state and determine the process used to get the bill passed. Examine the roles played by engineers or groups of engineers in the development and passage of the bill.

4. Develop a public policy statement that applies to some civil engineering problem on a college campus. This could relate to problems such as the accumulation and cleanup of trash in a local stream, a drainage problem, or a transportation/parking issue.

5. Obtain a copy of a state or country policy on storm water management. Adapt the policy statement to control of flood runoff on your campus.

6. Develop a policy statement that would increase the greening of your campus.

7. Draft a letter to the editor of a local newspaper about the proposed location of a waste treatment facility in your neighborhood. Use a limit of 125 words and write the letter for a nontechnical audience.

8. Draft a letter to a national news magazine (100-word limit) arguing that public monies spent on infrastructure are funds well spent.

9. Draft a letter to a federal agency chief who is using his or her personal bias based on the dictates of the executive branch to push for allowing more oil exploration on the North Slope of Alaska. Argue against the exploration.

10. Identify the recommended maximum length of letters to the editors of a local newspaper and a national news magazine (e.g., *Time* or *Newsweek*). Using recent issues of the paper and the magazine, determine the average word count of letters to the editor in each.

11. Obtain copies of the Clean Air Acts of 1963 and 1970. Analyze each on the basis of values and discuss the changes made. Hypothesize about the potential roles of civil engineers in fulfilling the recommendations of the two acts.

12. Locations with high traffic densities are subject to high noise levels, such as those caused by automobiles with bad mufflers or large trucks. What factors should be considered in developing legislative standards related to traffic noise? Briefly discuss each factor.

13. What are some social goals that have ties to civil engineering about which public policies would be relevant?

14. The goal of public policy at the federal level is to maximize the value of resources at the national level, possibly at the expense of the value to a region. Discuss the benefits and problems of this goal.

15. What are social costs with respect to public policy?

16. Using the sample storm water management policy of Section 9.5, evaluate the potential of the policy to influence both public and private expenditures.

9.8 GROUP ACTIVITIES

1. Table 9.1 includes a copy of the Ramsar Convention criteria related to wetlands of international importance. Analyze the criteria and discuss how each criterion could relate to the responsibilities of civil engineers.

2. Obtain a copy of a local ordinance on an issue relevant to civil engineering. Evaluate it from the perspective of values.

TABLE 9.1
Ramsar Convention Criteria for Identifying Wetlands of International Importance

Group A. Sites Containing Representative, Rare, or Unique Wetland Types

Criterion 1: A wetland should be considered internationally important if it contains a representative, rare, or unique example of a natural or near-natural wetland type found within the appropriate biogeographic region.

Group B. Sites of International Importance for Conserving Biological Diversity

Criteria based on species and ecological communities:

Criterion 2: A wetland should be considered internationally important if it supports vulnerable, endangered, or critically endangered species or threatened ecological communities.

Criterion 3: A wetland should be considered internationally important for maintaining the biological diversity of a particular biogeographic region.

Criterion 4: A wetland should be considered internationally important if it supports plant or animal species at a critical stage in their life cycles, or provides refuge during adverse conditions.

Specific criteria based on waterbirds:

Criterion 5: A wetland should be considered internationally important if it regularly supports 20,000 or more waterbirds.

Criterion 6: A wetland should be considered internationally important if it regularly supports 1% of the individuals in a population of one species or subspecies of waterbird.

Specific criteria based on fish:

Criterion 7: A wetland should be considered internationally important if it supports a significant proportion of indigenous fish subspecies, species, or families, life history stages, species interactions, and populations that are representative of wetland benefits or values and, thereby, contributes to global biological history.

Criterion 8: A wetland should be considered internationally important if it is an important source of food for fishes, spawning ground, nursery, or migration path on which fish stocks, either within the wetland or elsewhere, depend.

---

3. Discuss the roles of Congress and the executive branch in the development of federal policies. What conflicts develop in the process of finalizing a federal policy? Provide examples to illustrate points made.
4. Water resource policies are based on general principles such as ensuring sustainability, protecting the environment, minimizing the risk to the public from extreme rainfall and runoff, and providing clean water for public consumption. These policies form the basis for regulations and statutes. Identify and analyze a water policy related to one of these principles.

REFERENCES

Florman, S. C. 1976. *The existential pleasures of engineering.* New York: St. Martin's Press.
Tribus, M. 1978. The engineer and public policy making. *IEEE Spectrum* 47:48–51. April.

10 Globalization

CHAPTER OBJECTIVES
- Describe the globalization process and issues
- Discuss the meaning of professionalism in the context of global civil engineering practice
- Present criteria that can be used to address global issues

10.1 INTRODUCTION

Now more than ever, engineers must be aware that they impact more than just their local community—they impact the world and the events of the world impact them. Civil engineers in particular must be aware that their decisions and actions can have rippling effects on the environment, health, and lifestyle of people around the globe. We can attribute much of this newfound influence to globalization.

Globalization is generally viewed as a general issue; however, it can also be viewed from a civil engineering perspective because many issues related to globalization will require extensive involvement of civil engineers. These issues or problems include world health, environmental degradation and climate change, energy demand exceeding supply, building standards, structural solutions to the problems of terrorism, the distribution of food to the disadvantaged, and the dissemination of knowledge. Other issues could be cited. Worldwide solutions to such problems will require international collaboration and the overcoming of barriers such as language and cultural differences, nationalistic feelings and pride, differences in economic and political systems, and differences in business practices. If solutions to these global issues can be developed through cooperation of civil engineers in the affected countries, then global objectives will be met.

10.2 DEFINITIONS

Globalization is the integration over time and space of technical, economic, cultural, political, demographic, environmental, and social systems toward the solution of large- and small-scale problems, using ethically responsible standards of practice that are common across geographic and political boundaries. Globalization deals with issues using an integrated set of professional standards. From a globalization prospective, process, issues, and professionalism will be defined as follows:

Process: Actions taken and decisions made that facilitate greater cooperation across time and space while spanning political, cultural, and performance boundaries.

Issues: Events that have multinational, even international, consequences relevant to the natural, built, or social environments.

Professionalism: Overcoming differences in business practices, views on ethicality, standards of professionalism, including licensure, and cultural standards.

These concepts form the basis for achieving a global practice that meets the needs
of all societies and makes the most efficient use of global resources, including human,
environmental, and natural resources. The globalization process includes:

- Develop economic relationships between nations to integrate markets
- Provide for greater flexibility of labor movement
- Ensure that geographic boundaries are not constraints on the transfer
  of knowledge
- Provide for full utilization of information technology
- Remove international constraints such as language and culture barriers

Global issues include:

- Uniform world health standards
- Ensure preservation of nonrenewable natural resources
- Preserve the global environment
- Ensure engineering talent is properly distributed

Global professionalism implies:

- Having common ethical standards and respect for cultures
- Having uniform requirements for licensure
- Having uniform standards for permits

In the short term, globalization will not eliminate political boundaries, but those
in different countries will agree that everyone benefits when global goals are met
and global standards are applied to solve global problems. For example, air pollution
generated in one country does not stop at the geographic boundaries of that country.
Through transboundary crossings, air pollution can damage the forests and lakes of
other countries. Globalization is intended to encourage cooperation to reduce such
negative effects.

Globalization requires changes in the actions of individual nations, as the nations
cannot act solely in their own best interests. Globalization requires political differences to be reconciled in order for problems to be solved. Compromises will be
necessary to resolve multinational problems. Most importantly, for globalization to

succeed, individuals within countries will need to recognize the benefit of multi-national cooperation.

Globalization will require engineers to adopt different attitudes and will bring engineers from many countries closer together as they interact in the development, manufacture, and marketing of engineering solutions. This newfound global market for engineering services provides for greater exchange of technology, knowledge, and culture. However, an international market also creates a need for international licensing requirements, codes, standards, treaties, and trade agreements. Today, worldwide interactions are more than advantageous, they are essential; a country that seeks economic prosperity and technological progress cannot be sustainable under a closed economy.

Some developing countries realize that their economic growth requires a modern infrastructure, a clean environment, an educated workforce, and a knowledge base that may partially depend on foreign support. In many cases, those in developing countries may need the expertise from around the globe. This may include sending their citizens to other countries for their education. For some projects professionals from developed countries may need to temporarily relocate to the site of the development project. In other cases, professionals from a number of countries may need to work together by way of the Internet. These issues are all part of globalization and represent the spatiotemporal global integration of professionals.

10.3   VALUES RELEVANT TO GLOBALIZATION

Human values often vary from one culture to the next and in some cases over time. Therefore, globalization requires an acknowledgment of differing values and compromises among the participants to ensure that different standards do not cause neglect of the obligations that all participants have to public welfare, where public welfare now takes on a global perspective in addition to being a local issue. Beyond the common values of sustainability and public health, safety, and welfare, the changing nature of the following values is relevant to the globalization of engineering:

- *Allegiance*: Loyalty to a cause, nation, or institution.
- *Tolerance*: Respecting the nature, beliefs, or behavior of others.
- *Integrity*: Rigid adherence to a code of behavior.
- *Promptness*: Being punctual.
- *Knowledge*: Understanding gained through study or experience.
- *Efficiency*: The quality of producing effectively with a minimum of unnecessary effort or waste of resources.

Values are central to the success of any endeavor, including practice in a global workplace. For example, efficient globalization will require a change in allegiance from the nations of the participants to the project to the people who will be served by the project. Working across cultures will require greater tolerance and an expectation that the participants of the other cultures will not have to violate their own

CASE STUDY: THE WORLD'S FORESTS AS A GLOBAL ISSUE

Due to increasing worldwide demands of forest resources, we face many challenges to ensure effective forest governance. The social and ecological aspects of forest governance must both be addressed for effective policy. Public policy must carefully define user rights and responsibilities and encourage participation of those who utilize and depend on the forests. Engineers must vigilantly foresee and monitor forest outcomes, while officials and the community must constantly enforce and adhere to property laws (Agrawal et al., 2008).

Effective governance must involve viable policy that includes input from technical experts, such as civil engineers. In turn, civil engineers must be aware of the needs of the people and consider the needs and demands of the communities that directly depend on forests for both their livelihoods and recreational activities. Civil engineers must be aware of the global impacts of environmental actions. It may be difficult, at first, to understand how the choices that people in other countries make can impact the United States, and vice versa. The idea of the "butterfly effect" is one way to view this impact. As the theory goes, when a butterfly in the Himalaya Mountains flaps its wings, the effect is felt worldwide. Specifically, a small initial change can cause a chain of events that ultimately leads to a large-scale alteration of events. For example, slash-and-burn techniques or hazardous waste dumping can harm the natural resources of the atmosphere and water for people halfway around the globe. Deforestation of land for use as biofuel, food plantations, or timber sales can advance erosion in the local area and also affect global climate change.

Ethics and rights come into play when engineers enter the field of environmental decisions and policies. Engineers must ask themselves questions, many of which do not have clear solutions. For example, is it fair for a neighboring country to intervene in the Amazon's rainforest timber industry in Brazil to save plant species that may one day provide cancer-fighting medicine? If so, is it ethical to reduce the profits of one community or country in order to provide benefits to the global community? Is it fair for indigenous island inhabitants to have their water and food supply ruined by oil spills caused by other countries' need for energy? Is it ethical for a country to intervene in another country's decisions if that country is making environmental decisions that are negatively affecting the world health? Answers to such questions are difficult to provide, as each country approaches the issue from a different perspective.

The number and degree of connections that countries share with one another is constantly increasing. Boundaries that determine authority are becoming less defined, as environmental effects have worldwide consequences. The web that we call globalization is affected by each decision that we make: symbolically, a slight tug on one end of a string can cause the whole web to change.

beliefs and adopt the ways of those from the lead nation. Behavior varies from one culture to another. Therefore, a common understanding of what constitutes integrity will need to be developed. Where face-to-face communication is difficult, such as when participants are in different time zones, promptness will be more important. Engineering needs and experiences differ across cultures, which will lead to different knowledge bases of the participants. These could create conflict if the participants are not tolerant of the differences. A waste of the resources of one group should not be acceptable to participants from other countries, as resources must be viewed from a global perspective.

10.4 CRITERIA TO ADDRESS GLOBAL ISSUES

Civil engineering projects often require considerable use of natural resources and have significant impacts on the local environment. For example, large wood structures require the use of forest resources. Also, construction projects often discharge sediment to receiving streams, and illegal dumping of waste construction materials into streams is not uncommon. While these may seem to be local issues, they may have regional or global implications, and therefore criteria are needed to evaluate the projects from a global perspective.

While the civil engineering perspective on global issues is of direct interest, general criteria related to global issues cannot be ignored. The importance of one criterion relative to another criterion will, of course, depend on the issue, so the evaluation of the criteria needs to be specific to the issue. The following criteria should be considered when dealing with a global issue:

- Natural and human resources of each political unit have been used wisely.
- Global sustainability measures have been met.
- The project benefited each political/economic unit and possibly other non-participating units.
- Cultural differences were addressed to the satisfaction of each unit.
- The concerns of all stakeholders were considered in an unbiased way.
- Agreement was reached on ethical issues, with all differences considered.
- Business goals were met.

Other criteria relevant to a specific global issue could be added to the list.

Each of the above criteria is intended to encourage engineers to apply a global perspective to their projects. These criteria discourage selfish criteria, such as the use of resources in a way that maximizes project efficiency, and encourage the adaption of global criteria. Would a civil engineering company with a project in another country sacrifice profits in order to provide for sustainability of a host's natural resources? If globalization is to be successful, at least with respect to engineering, companies and individual engineers will have to be willing to apply global criteria to their practice, possibly even replacing common business practices like maximizing profits with global criteria. This will reflect a significant change in attitude.

CASE STUDY: COAL PLANTS IN CHINA

The design, construction, and use of a new coal-burning electrical generator plant can cause the discharge of pollutants into the atmosphere, which can be distributed worldwide. This is true whether the plant is in Topeka, Kansas, or Apia, Samoa. While the direct users of the electricity will benefit from the power plant, many who will not benefit will be subject to the pollutants.

China is just one example of a nation that relies on coal to provide the energy needed to advance its economy and social systems. China's increasing use of coal is mainly due to the improving living conditions of many Chinese people. Second, China possesses very few reserves of natural gas and oil, so coal is essentially the only energy option available. However, burning coal to produce electricity and run factories generates more greenhouse gases and lung-damaging pollutants than burning oil or gas.

Researchers in California, Oregon, and Washington found by-products of coal combustion, including sulfur compounds and carbon, coating their mountain-top detectors. These microscopic particles can cause cancer, respiratory disease, and heart disease. In the mountains of eastern California near Lake Tahoe, filters were the darkest that the atmospheric scientists had ever seen, and the researchers believed that this was due to the air transport of the particles from China. Pollution from Asian coal-burning plants is already making it more difficult and expensive for West Coast cities to meet their air quality standards. The Chinese government is vigilantly working to decrease emissions and now requires that the smokestacks of new coal plants be outfitted with devices that remove up to 95% of sulfur emissions. The government plans to install these devices on all existing plants by 2010. Americans should not be quick to assume that China is the source of our emissions dilemma. International climate experts point out that the average American consumes more energy and is responsible for ten times more carbon dioxide emissions than the average Chinese person. While China does generate more electricity from coal than the United States, our gasoline consumption is larger (Barboza and Bradsher, 2006).

An obligation that a country has to its own society often conflicts with its obligation to the global society. Chinese leaders are under pressure to improve the quality of life and create jobs and opportunities for millions of people who are coming into the cities from the countryside seeking employment. Is China's use of coal justifiable? The developed countries went through their industrial revolution, so isn't it fair for China to have its revolution? This situation is a prime example of the interconnectedness that we share with other nations and the ethical issues that we encounter.

We must consider the realms of engineering ethics, economics, and environmental stewardship in today's era of globalization. This issue is certainly not confined to the effects of power plants in China. For decades, Canadians have blamed much of their acid rain problem on power plants located in the United States. The point is: many significant problems are global reaching in nature, and solutions will depend on engineers from many countries working together. The problems affect the engineering profession, and a global engineering attitude will be needed to solve these global problems.

10.5 THE GLOBAL ENGINEERING WORKPLACE

10.5.1 LEVELS OF GLOBALIZATION

According to Friedman (2006), the word *globalization* can be described by three distinct phases. The first phase involves *countries* globalizing. The second phase involves *companies* globalizing. The third and present phase involves *individuals* globalizing while competing and collaborating around the world. While the first two phases were primarily initiated by European and American businesses and individuals, a much broader array of nations are now participating. The third phase is run by individuals in greater quantities from all areas of the globe. This third phase of globalization, which has been increasingly growing over the past ten to fifteen years, represents the cultural and economic aspirations of hundreds of millions of people around the world. The effect of globalization is exponential, and as more people and economies get involved, its speed and range of influence will increase.

10.5.2 THE GLOBAL ENGINEERING WORKPLACE

An initial thought might suggest that the global engineering workplace will be vastly different from the traditional engineering workplace at the corner of Main Street and First Avenue. The global workplace will replace a meeting in the office conference room with the Internet as the center of activity. The client of the global workplace may never meet face-to-face with the project manager. The design team will not all be graduates of the local state university, but graduates of universities around the world, such as the Indian Institute of Technology, Kansas State University, and Oxford. Instead of only meeting the building standards of cities such as Bozeman, Montana, the design team will be addressing global building standards that meet the required safety criteria of all participating countries. These are just a few of the changes that can be expected as progress is made toward a global engineering workplace. However, the design team will still provide a safe design that meets the specifications of the client—this will not change.

10.5.3 GLOBALIZATION AND THE JOB MARKET

Globalization is changing the field of civil engineering in monumental ways. Although many domestic opportunities are available, American civil engineers are now competing for jobs with engineers from overseas. As a 2008 *Science and Engineering Indicators Report* by the National Science Board found, the percentage of overseas students pursuing an undergraduate engineering degree is much higher than that in the United States. In Asia, 20% of students seeking their undergraduate degree are engineering majors, while many other countries have more than 10% of their undergraduate students in engineering. In the United States, only about 5% of undergraduate students are engineering majors.

According to the Bureau of Labor Statistics' *Occupational Outlook Handbook*, many well-trained English-speaking engineers are available overseas and are willing to do the same work for much lower salaries than American engineers

GLOBALIZATION: BURJ KHALIFA

The design of a mega-story structure of Dubai involves engineers from around the globe. They collaborated on the design work in real time even though each group was working in their own city.

According to Reina and Post (2006), Burj Khalifa, called Burj Dubai before its inauguraton, was projected to be the world's tallest building. Burj Khalifa is a considerable feat in civil engineering and architecture, at 2716.5 ft and more than 160 stories. As of October 4, 2010, it is the tallest building in the world.

The construction of Burj Khalifa involved companies from the Middle East, Germany, the United States, Australia, South Korea, Belgium, and the United Arab Emirates. The design and construction required cooperation from architects, suppliers, construction companies, and engineering firms from around the globe. The tower's architect is from the United States. Also working on the design are companies from the United States and Canada. Multicountry experts are working together and using wind data to optimize the orientation and shape of the tower. Building safety and fire codes were addressed as well, using several international building codes.

The 2,500 tons of formwork for the core and walls of the tower were constructed by an Austrian company, and the two high-pressure pumps used to deliver the concrete were from a German company. Additional partnerships include materials from Australia, the Czech Republic, and Switzerland (Reina and Post, 2006). Many of the construction workers and laborers are migrants from China, Bangladesh, India, and Pakistan who are in Dubai solely for construction opportunities, for projects such as Burj Khalifa (Krane, 2006).

This construction project is an example of many projects where international teamwork is relevant around the globe, especially for large-scale construction projects. With technology that allows real-time responses, such as teleconferencing and personal device assistants that can be used at the work site, the engineers and workers of Burj Khalifa can be interconnected at all times. Most importantly, the project indicates that globalization can work.

(U.S. Department of Labor, 2007). The advancement of the Internet and other technologies is making it simple for engineering work that previously had to be done domestically to be completed in other countries, with all communication completed by telephone and the Internet.

Even though an ample number of opportunities exist for civil engineers in any one country, engineers must be aware of the overseas competition and possible job outsourcing. Outsourcing is the subcontracting of a service, such as a project design or manufacturing activity, to a third-party company. This is believed to improve the efficiency and profits of a business. Routine work and calculations are often outsourced so that more advanced work can be done domestically. American engineers must be trained to work at the management level and to aim for the more advanced work that overseas firms may outsource to the United States. Raising the standards